Plant Propagation

D0531377

Peter Klock

Plant

Propagation
House and garden plants

WARD LOCK

Contents

Preface

We have already put the throw-away society behind us, and the mood of the times is for health and wholeness, and those things in general which promote life. Those things, in fact, which go back to nature itself. That is not to say that there are not artificial inventions in our lives today which we cannot imagine losing. But wherever possible, we are tending more and more to go back to natural products. Growing our own fruit and vegetables is an example of this. Many people are increasingly coming to appreciate having flowers they have grown themselves in their gardens and their homes.

Successful plant growing calls for more than rushing straight in to raise plants, with nothing but good intentions to guide us. We can hardly expect lasting success without some knowledge of the "rules" of nature. This book deals with questions about the propagation of plants. There are questions which come back time and again: what must I look out for when buying seeds? How can I avoid buying the wrong thing? How should I sow the seeds, and when is the best time? What sort of soil do I need, and what temperature is necessary? Why do fruit trees need to be grafted? When is the right time to take cuttings? What do I do about fungal diseases in my cold frame? How can I raise new plants from my *Fatsia japonica*? Is it possible to cultivate Bonsai oneself?

This book aims to give answers to all these questions. It shows how plants are propagated, what different methods and possibilities exist, and what in particular to look out for.

Propagating plants yourself can sometimes save you money, but above all it brings great pleasure and satisfaction to watch plants you have propagated yourself grow into prize specimens.

The advice and tips given in this book are founded in a wealth of experience of horticultural practice. Nevertheless, success cannot be guaranteed. Again and again, plants will develop differently, or even not at all. They are, after all, living creatures.

Working with plants demands care, conscientiousness and patience. Another piece of advice: some species of plant can cause injury; some are even poisonous. That is also true of some fruits and some of the seeds named. You should always take care when working with them. Small children in particular should not be allowed to handle plants unsupervised. But it is important for children especially to see the miracle of plant growth. This book therefore also contains a chapter giving suggestions for children to propagate plants.

Germination of the cucumber

Peter Klock

A few words about botany

When does a plant begin its life? With fertilization, and the formation of the seed which results? Or when the seed germinates? Or somewhere in between?

Propagation from seeds

The life of a plant really begins as soon as fertilization takes place. Only then can the seed develop, and it contains all the features of the plant it will become. It is simply dormant, and needs to be woken from this sleep by giving it the right conditions to germinate.

A seed basically consists of three parts, the embryo, also called the **seedling**, the **endosperm** and the **seed coat**. The endosperm contains the food reserves for the new plant, and in grain and grass seeds (rye, barley, oats, rice and bamboo) this mainly consists of starch. In leguminous plants (peas, coral-tree and cassia) the reserves consist of proteins, and in oil-seeds (olives and jojoba) they consist of fats. The embryo usually lies directly in the endosperm. Sometimes food reserves are also stored in the seed leaves (cotyledons), as in the pea. If the seedling has one cotyledon, it is a **monocotyledonous plant** (Monocotyledoneae). This includes grasses (Graminae), the lily family (Liliaceae), the iris family (Iridaceae), bromeliads (Bromeliaceae), palms (Palmae), orchids (Orchidaceae) and the banana family (Musaceae). If there are two cotyledons, it is a **dicotyledonous plant** (Dicotyledoneae). These include amongst others the deciduous shrubs and cactuses (Cactaceae). Conifers, which belong to the gymnosperms, usually have several cotyledons. They are called **polycotyledonous plants**. Germination is

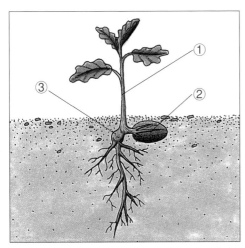

Hypogeal germination: ①*Epicotyl with foliage leaves* ② *Cotyledons with food reserves* ③ *Hypocotyl*

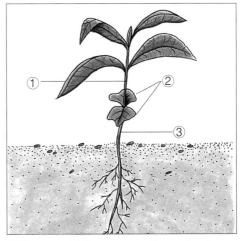

Epigeal (above-ground) germination: ①*Epicotyl with foliage leaves* ② *Cotyledons with food reserves* ③*Hypocotyl*

divided into two types, below ground, known as hypogeal germination. Above ground, this is known as epigeal germination.

Plants which germinate below ground include, for example, the oak (*Quercus*). The thick seed leaves or cotyledons, containing the food reserves, and the **hypocotyl** (the germinating shoot of the seed-bearing plant, the region of the stem forming the transition zone between the root and the shoot) remain in the ground. The seedling, consisting of the **epicotyl** (leafless section of the shoot) and the foliage leaves, is above ground. An example of a plant which germinates above ground is the coffee plant (*Coffea*). Here the hypocotyl is above ground, then come the cotyledons, and above those the epicotyl and the foliage leaves.

Propagation from parts of the plant

Plants are not only propagated generatively, or sexually, from seeds. They are also often propagated vegetatively, or asexually, from parts of the parent plant. Since this is done using only cells from the parent plant, all plants propagated in this way are genetically identical.

Various herbaceous perennials form **rhizomes** (underground stems used by the plant for storage). These enable them to survive without any plant parts above ground after they have died down, and they are dormant between one growing season and the next. If the rhizomes are artificially divided, they can form new shoots and so new plants. This makes garden weeds which form rhizomes,

like couch grass or bindweed, very difficult to combat.

Other organs of vegetative reproduction are **bulbs**, which consist of a shortened shoot that has thickened leaf bases, in which food is stored. There are also **corms** and **tubers**. These are subdivided into stem tubers, such as the potato, in which the end of a shoot is thickened, and root tubers, such as the dahlia, which has growth buds at the crown only. It is important to be aware of this when planting.

Finally there are beets, which consist of the thickened shoot and root. These give up their stored reserves at an early stage to form the flower head.

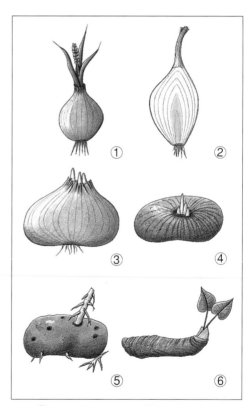

① *Bulb (sprouting)* ② *Onion (cross-section)* ③ *Corm (crocus)* ④ *Root tuber (anemone)* ⑤ *Stem tuber (potato)* ⑥ *Rhizome (arum)*

The greenhouse, the cold frame and indoors

The greenhouse

One of the greatest pleasures for plant-lovers must be to have the use of a greenhouse, or even to own one. In it they can raise for themselves basically all the plants they are interested in, from seed or vegetatively. But anyone who is thinking of acquiring a greenhouse should consider before buying whether he or she does really need one, as they can often be quite expensive. Once the decision has been made to buy a greenhouse, the next questions to be answered are what should it be made of? How big should it be? And what sort of roofing should it have? Of course it may not be possible to have everything you want. The neighbours' wishes need to be taken into consideration, the building must be suited to the type of ground, it must be possible to provide heating and a connection to a fuel supply (gas, heating-oil tank, supply from the house, electricity and water supplies), and all at a manageable cost. Some compromise is inevitable. But it is a mistake to do without the features you need to use the greenhouse for the purposes you really want. For example, anyone who is interested in having the facility for raising plants at any time ought not to do without the means of heating the greenhouse.

The decision as to the size of the greenhouse is particularly important.

When everything has been taken into account, and you know what size you want it to be, it is a good idea to make it just a little bigger. Greenhouses so often turn out after a little while to be too small, normally when plants need to be overwintered. It is very bad for them to be packed together too closely in winter. If the ventilation is not working very well – often a problem in small greenhouses when the weather is very cold outside – the way is wide open for the spread of pests and diseases. Fungal attacks can be especially bad and difficult to bring under control in these circumstances.

When deciding the size of the greenhouse, you need to consider whether heating is necessary, and what the minimum temperature in winter should be. The smaller the greenhouse, the greater the percentage of glass to its volume, so the greater the relative energy requirement will be. The type of roofing is important here. Should the greenhouse

The greenhouse should be large enough to be able to overwinter plants

9

be covered cheaply with horticultural sheeting, or should it be glass, or even twin or triple-walled panels?

If the greenhouse is to be heated, ordinary horticultural sheeting is not advisable. Tunnel greenhouses covered with plastic film which has extra thermic properties and anti-fogging properties are an exception. There may or may not be an additional insulating layer of bubble film. Otherwise, houses with glass roofs made of plain or treated horticultural glass (which gives a diffuse light) are preferable. It is still better to double-glaze, or to add insulation by using sheeting such as plastic bubble film, although this reduces the light which can get in. Twin or triple-walled polycarbonate panels give very good insulation, but these materials, along with insulating glass covering, are amongst the most expensive. This

example, however, makes the point: spending on insulation saves substantially on energy running costs. Spending little means paying more for heating. A lean-to greenhouse is good from the point of view of energy costs, because at least one side can benefit from the warmth from the house wall. It also has some visual appeal, in bringing the conservatory to the door. Connection and energy costs can be kept much lower if the heating system can be linked up with the one in the house.

Triple-walled panels (section shown here) provide good insulation

A lean-to greenhouse helps to save energy

Another consideration in building a greenhouse for raising plants is the need for shading from strong sunlight. Various systems are available from specialist suppliers. Mounting a roll of matting on the roof, to be rolled up and down as needed, is simple and effective. Shade can then be provided when necessary. Whitewashing the panes is common, but the shade provided is permanent and it does not look very attractive. It is therefore not recommended.

Finally, a word or two about equipment. Having a workbench makes life much easier. Benches made up of aluminium sections or galvanized steel are good; they will be less affected by the damp, warm atmosphere. Stone or concrete benches could also be used. Wooden or iron benches are not advisable, even if they are painted. They involve too much maintenance. In the past, asbestos cement was used very effectively for the work surface itself. This went out of use for some time

Greenhouse workbench

because of the asbestos content, although an asbestos-free version has meanwhile come into production. If the work surface is made up of panels which are not too large, wire-reinforced glass is also a possibility. The advantage is that plants placed underneath do receive some light (but beware of sharp edges).

Heating

A greenhouse which is used for raising and cultivating young plants should have heating. Only then can it be used to the full. This is because a number of propagation techniques are mainly carried out in winter. If you wish to cultivate tropical plants – most well-known house plants come into this category – the heating must be adequate to provide a minimum temperature of 15^0C (59^0F) in winter.

The most practical way of heating a greenhouse or conservatory is to run the supply directly from the house. It is particularly helpful if the glasshouse is very near or built on to the house. A **hot-water central heating system** is ideal. Heat is given off continually by con-vection, without overheating any part of the space too much. An aeration fan can be installed to distribute heat quickly and evenly through the greenhouse.

Commercial establishments raising young plants often favour **floor heating systems**, if they have hot-water heating. Heat is distributed by way of plastic pipes, which are laid in the ground at a depth of about 30cm (12in). Polystyrene panels should be laid underneath them for insulation. This sort of heating system warms the ground and the space

11

Electric heaters must be specially equipped for use in the greenhouse

Gas heating is an economical solution for greenhouse owners with access to a gas supply

above it very well. Thermostatic regulation is possible, but it is fairly slow to react.

If gas is available, and if the intention is to heat the greenhouse directly, then **warm-air heating** is a reasonably priced alternative. The system is set up in the greenhouse and the flue led outside. Thermostatic regulation ensures that the desired temperature is maintained constantly. Connection to the electricity supply is necessary. A gas warm-air system set up in the greenhouse, but without a flue led outside, is cheaper. It can also be thermostatically regulated. But this does have the disadvantage that the oxygen in the greenhouse is used up, and the carbon dioxide level is considerably raised. This can be a

particularly serious problem if the greenhouse is well insulated, and not enough fresh air can get in. One possible result of this is that the flame can go out. Also, water is formed in the combustion process, and this collects on the underneath of the roof glazing. It then drips down if there is any draught of wind or movement. Spores of fungus can be spread this way, and stimulated to grow. **Electrical heating** is very clean, but also very expensive. Some manufacturers of small greenhouses offer this as an accessory. Convector heaters in the form of a tube with convection surfaces, mounted near the glass towards the bottom of the greenhouse, can be recommended. This positioning prevents the glass misting up. Electrical warm-air fan heaters can also do a good job in a small greenhouse. But only specially designed and insulated models approved for such purposes should be used. They cost several times as much as the ones designed for domestic use. Thermostatic control is recommended. This brings big cost savings, and means that the right temperature for the plants can be set and maintained. The installation costs for electrical heating are usually much lower than for other heating systems, although the electricity consumption costs tend to mount up.

Additional lighting

Light is essential for plants to survive. It is their source of energy, and they cannot live without it. Plants convert simple inorganic substances into complex organic ones. This process is called **photosynthesis.** In it, the water which the plants take from the ground and the carbon dioxide (CO_2) they obtain from the air are processed into starch and oxygen with the aid of sunlight. The oxygen is produced as a waste substance by plants, and is continually being given off into the atmosphere. Human beings and animals need oxygen in order to survive. So without plants, there could be no human or animal life on earth. Photosynthesis happens in the presence of chlorophyll, which gives leaves their green colour. Many native trees lose their leaves in autumn. They then become dormant, of necessity. Certain plants without chlorophyll, like red or yellow

Red cacti cannot survive without a rootstock

Effect of raising seedlings with insufficient light: the plants in the rear tray are drawn up; those in the front tray show normal growth

cacti, cannot survive on their own for the reason just explained. They therefore have to be grafted on a rootstock.

Raising plants in winter is certainly possible, and for many species it is recommended. A greenhouse or indoor propagator is particularly suitable for this purpose. There the plants have the temperatures they need to emerge and continue to grow. But bear in mind the fact that most species raised in this way do not come from temperate parts of the world, but from the subtropics, or even the tropics. The temperatures there are similar to those in the greenhouse, but the amount of light is not. So the plants suffer from lack of light, especially in winter. They become etiolated or drawn up. That is, they form long, soft, straggly growth with large internodes (this is the name for the stretch of stem in between the buds). Beyond a certain length they cannot support themselves and fall over. It is then simply a question of time before they rot or are attacked by ground fungi and die.

It is important, then, to ensure enough light. Various firms offer lighting for the purpose. The special lighting

13

tubes which approximate to daylight (they give a bluish-violet light) are one particularly inexpensive but effective option. They are available in various wattages and are installed like ordinary lighting tubes.

However, only the version approved for use in horticulture, which has protection against water splashes, should be used.

From mid to late spring, additional lighting for the raising of young plants is usually no longer necessary.

For indoor propagators, there are special arrangements consisting of a heating element, a plastic seed tray to go with it, a transparent cover and a frame above it with a built-in fluorescent tube. These usually give off white light, however, because it is pleasanter on the eye. These fluorescent tubes are admittedly not quite as good as the ones mentioned above for providing plants with supplementary lighting.

With such supplementary lighting, a great range of plants can be grown, even in a fairly dark corner of a room, without worrying about the plants becoming etiolated and dying.

If the temperature of the heated base can be thermostatically controlled, and the light has a timer to turn it on and off, this of course means that the propagator will require still less attention.

The length of time for which the artificial lighting should be switched on each day is about ten to twelve hours.

Mist propagation

Mist propagation was developed at the end of the 1930s in the USA. The intention of this method is to make it

Special lighting for greenhouses

Correct level of lighting

A rule of thumb for deciding how much light is needed is about 50 watts per square metre at a height of about 30cm (12in) above the plants being lit. The plants should however have a light intensity around 2,000 lux at least. Depending on the particular lighting equipment chosen, this would involve installing a 40-watt daylight lighting tube about 1m (3ft) above the plants.

easier to raise cuttings by automatically keeping the leaves moist, to keep the pressure of the sap in the cutting constant and so prevent it from wilting.

A mist propagator basically works like this: water is sprayed through very fine nozzles inside an enclosed propagation unit to form a mist. This creates 100 percent relative humidity. Evaporation of water is the main enemy of young unrooted cuttings, and this method prevents it.

If they are not losing water, the leaves do not need to be removed from the cuttings, but can be left on to carry on photosynthesis. That promotes rooting, and completes the process: cuttings planted under mist propagation root much more quickly. Softer cuttings can also be used, because they remain turgid under mist (the pressure of their sap is maintained). An additional benefit is that fungal infections and attack by pests hardly ever become a problem. Using mist propagation, it is possible to root even plants which do not normally take well from cuttings.

There are some disadvantages to mist propagation, though. It only works when there is water pressure of at least 3 bar. Soft water is needed, otherwise the system will fur up and stop working properly. The length of time for which the spray operates has to be controlled. This is usually done using a so-called electronic leaf; the weight of the film of water on a metal plate operates a switch mechanism. This prevents the ground becoming soaked, and the cuttings rotting. Evaporation of the water in the spray causes cooling. Bottom heat should therefore be used to ensure favourable temperatures.

Mist propagators can also be installed in smaller greenhouses, if the practical demands mentioned can be met. Any malfunction of the unit needs to be dealt with straight away, to prevent the plants from being damaged.

The cold frame

Cold frames were once viewed as an

Cold frame

important, even essential aid in the raising of young plants. In the meantime, plastic tunnels, garden fleece and other similar products have become available very cheaply, and fulfil the same purpose. Nevertheless, cold frames do still have their use, as they are very suitable for cultivating cuttings which are to be planted out in mid or late spring. They are excellent for raising and forcing various summer-flowering plants, and tender vegetables like cucumbers, capsicums and aubergines. These frames are ideal for providing a constant supply of fresh herbs and spices, too. Here they are protected and can grow well; they also have ideal conditions to grow back after cutting. A further use is for propagating conifer cuttings.

Raising plants indoors

A great many plants can be grown from seed in a container on the windowsill. Having no garden or balcony need not mean that you cannot grow plants. Most of our houseplants come from warmer regions, but so do many of our garden

15

Propagator for the windowsill

plants. So they are very well suited to home cultivation on the windowsill. Some are decorative in themselves, some have flowers, and some even produce edible fruits.

Even flowering and fruiting plants you have raised yourself, like the following, flourish very well in the home: abutilon (*Abutilon*), allamanda (*Allamanda*), the pineapple plant (*Ananas comosus*), various begonias (*Begonia*), calamondin orange (*Citrofortunella mitis*) cotton (*Gossypium*), oleander (*Nerium*), species of passion-flower (*Passiflora*) and many different bulbs and corms.

A number of seeds of exotic fruits can be made to germinate in an indoor propagator. However, the plants which result often grow so vigorously that they need to be transferred to a greenhouse, or even outside into the garden in summer. Here is an interesting possibility: it is possible to grow miniature plants from the seedlings of potentially normally sized plants or trees. Small root baskets are available, consisting of woven mesh. These are filled with propagating compost and put

into Jiffy pots. The seeds sown in them grow into small plants which can develop only a limited root system, and so remain small. At the same time the mesh allows through the water and nutrients needed by the plant.

Adequate warmth is almost always present in the home, but there is often a lack of light, especially if the pots or trays are not immediately next to a window, but further back in the room. Then supplementary lighting will be needed (see page 13).

Certain seeds need a cold winter period to overcome their state of dormancy, so that the seed can germinate and grow. Many trees and shrubs are in this category. It is better to raise these out in the open, or in frames.

Plants which can be raised indoors from seed or cuttings

Foliage plants such as the dragon tree (*Dracaena*), yucca (*Yucca*), various species of fig (*Ficus benjamina* and *Ficus elastica*, the rubber plant), the banana tree (*Musa*), the dieffenbachia (*Dieffenbachia*), also the spider plant (*Chlorophytum*) and a number of species of palm and cactus. The coffee plant (*Coffea*) can also bear blossom and fruit.

The propagating medium

Cuttings are rooted in a propagating medium. Usually this is a seed or potting compost. Some species such as the willow (*Salix*), forsythia (*Forsythia*) and the oleander (*Nerium*), even root without difficulty in water.

It is possible to prepare propagating composts yourself. A simple recipe is one part not too fine clean or washed quartz sand (building sand) and one part fine-fibred peat. Fine sand would tend to clog, or be washed out on watering. The proportions can be varied considerably. Many professionals have their own mixtures, and some even keep these strictly secret, especially as regards the additional ingredients.

If the peat content is high, or if peat alone is used for cuttings, it should not be too acid. If necessary, lime should be added to bring the pH up to 5–6.5.

The following points should be attended to when mixing propagating media:

● The compost must be capable of holding moisture, but not tend to clog: cuttings in wet compost cannot breathe, and so rot very quickly.

They are weakened and so more likely to fall victim to attack by pests or fungal infections.

● No nutrients should be added to the mixture: only in a medium which is poor in nutrients will the young plant or seedling engage itself intensively in building up a good root system to provide itself with nutrients.

● The medium should be free of harmful organisms and weed seeds: if these are present, it could cause considerable damage to the tender young plants, or even kill them.

Self-mixed soil-containing composts should therefore be sterilized by heating them for an adequate time to a sufficient temperature in suitable heat-resistant wrapping. The simplest and safest solution is to buy good proprietary composts. These types are prepared specifically to meet the needs of seedlings and young cuttings, and are free of pests and weed seeds.

Pomegranate cuttings in propagating compost

Soil sterilization

Heat 3–5 litres of self-mixed soil-containing compost for half an hour in a domestic oven at 140–160°C (284–320°F). This will reliably kill harmful organisms.

Aids to rooting

Certain cuttings are difficult to root. Some species are helped by mist propagation (see page 14). Often, it is helpful to use substances which promote growth and noticeably encourage rooting. These growth substances are available to buy. The substances involved are any of a number of chemical compounds (hormones) which are particularly active on softwood, hardwood or semi-hardwood stem cuttings. Their effect is different according to species.

They are easy to use, as most of these preparations come in powder form. Before the cutting is put in, the end is quickly dipped into the rooting powder, the excess tapped off against the rim of the container, and the cutting is put into the soil. Some of the preparations have to be dissolved in water, and the cuttings are placed for a while in the solution. The soil temperature should not be allowed to rise beyond a certain temperature, of around 24⁰C (75⁰F), while the cuttings are being raised. The instructions accompanying the rooting preparation contain further details and these should be read before use, and followed carefully.

Giving the names of particular products is not all that helpful, because – as with pesticides – by the time this book appears, some may no longer be approved for use, or may have been replaced by others.

Cuttings are dipped in rooting powder before being put in

Cuttings from the following plants, amongst others, root particularly well with the use of rooting hormones: Species of maple (*Acer*), camellias (*Camellia*), Japanese quince (*Chaenomeles*), citrus species (*Citrus*), clematis, traveller's joy (*Clematis*), witch hazel (*Hamamelis*), hibiscus (*Hibiscus*), types of apple (*Malus*), mulberry (*Morus nigra*), olive (*Olea*), firethorn (*Pyracantha*), rhododendrons and azaleas (*Rhododendron*), gooseberry (*Ribes*), highbush blueberry (*Vaccinium corymbosum*), snowball tree (*Viburnum*), weigela (*Weigela*) and jujube (*Ziziphus*).

18

Propagation by seed

Most plants can be propagated by seed. In contrast to vegetative propagation, in which the new plants exhibit only the characteristics of the single parent plant, the plants which result from sexual reproduction have characteristics inherited from both parents. After pollen has been deposited from the male organs of the flower (the stamens) on to the female stigma, and fertilization has taken place, seeds are formed.

A number of plants are **self-pollinating.** In these species, the male and female organs are usually situated within the same flower. Visiting insects or movement caused by the wind transfer the male pollen to the female stigma. Then, when fertilization has taken place, the seeds can form. Plants of this type include beans (*Phaseolus* sp.), tomatoes (*Lycopersicon*) and various species of citrus (*Citrus* sp.). Fertilization by pollen from another plant is of course also a possibility.

Another variation consists of plants which are still self-pollinating, but which carry separate **male** and **female flowers** on the same plant. Here, pollination occurs when pollen is transferred from the male to the female flower. Plants of this type include cucumbers (*Cucumis* sp.), pumpkins (*Cucurbita* sp.) and even one variety of kiwi (*Actinidia*). Pollination from another plant is of course possible here too.

Finally there are whole plants which are either **male** or **female.** They must always be planted at a certain distance from each other, to ensure that pollination can take place. The distance apart depends on a variety of factors. Most of these plants rely on insects to pollinate them. Examples of these plants include the sea buckthorn (*Hippophae rhamnoides*), kiwi (*Actinidia chinensis*), skimmia (*Skimmia japonica*), the mastic tree (*Pistacia lentiscus*) and spinach (*Spinacia oleracea*).

There are some further variations. Amongst sweet cherries (*Prunus*) and the avocado (*Persea americana*), for example, only the pollen from plants belonging to quite particular groups can fertilize the female flowers of another. In the case of the walnut (*Juglans regia*), male and female flowers occur on the same tree, but they mature at different times, so that self-pollination is usually impossible. Cross-pollination gives a mixture of inherited characteristics, and for many plants this is essential to their survival. Many tree species, for example, can only adapt in this way to the frequent changes in environmental pressures. This is how they prevent

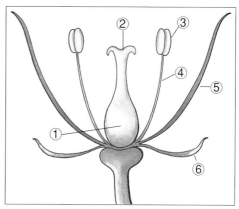

Schematic structure of a flower:

① *Ovary* ② *Stigma* ③ *Anthers*
④ *Filaments* ⑤ *Petals* ⑥ *Sepals*

themselves from being damaged or dying out as a result of a changing environment. Natural selection ensures that only the strongest and best adapted individuals are perpetuated. Diseases can still cause great damage, but plants are constantly emerging which are better able to survive them. Only these plants are then able to reproduce and so pass on their resistance to the next generation. This mechanism has enabled plants to survive even the most difficult conditions.

The genetic variety of the plant world is thus indispensable. This does not mean that there is any contradiction here with vegetative propagation, as long as we do not use it to raise vast monocultures, but simply to propagate plants to meet immediate requirements. Some plants multiply vegetatively quite naturally in any case, without any human intervention, for example by putting out runners.

Buying seeds

Seeds are available almost all the time for a great number of plants. They are on sale from specialist and many other stockists. Special mail-order suppliers often either specialize in particular seeds or offer a large range of specialties. There are strict regulations governing many vegetable and agricultural seeds. They should have a certain germination capacity, and – perhaps obviously – they must be true to type.

The sell-by date printed on many seed packets is very helpful. It only applies of course if the seeds are stored properly in the original packaging. Some seed packets now also show the number of plants which can be expected to result. Seeds in packaging which offers germination protection are protected against variations in climatic conditions. The high humidity which can harm dried seeds will have no effect on these.

Specialist mail-order suppliers of seeds often pack the seeds individually to send to their customers. These are often seeds of exotic plants, imported from every country under the sun. It should be noted that the germination capacity of such seeds can be well below that of home-produced seeds.

Germination testing

To test the viability of a particular lot of seed, a germination test can be carried out. This gives the percentage of the particular seed which germinates within

Storage of seeds

● Follow the instructions on the seed packets.
● For most species, storage should be in cool, dry conditions.
● Use exotic seeds as soon as possible.

a given time. The extent to which it is possible to predict on the basis of the test depends on a number of factors. It is only useful if the parameters of the test are chosen sensibly.

When is there value in performing a germination test, and when not? Germination tests on old seeds will tell the gardener how thickly to sow them. Quality control can be carried out on the seeds purchased, if the normal germination capacity is known. The germination capacity is important in agriculture and knowing the figures enables decisions to be made about whether to undertake the expense of buying more seed.

It makes less sense for the amateur gardener to make germination tests on exotic seeds, because they are expensive, and priced individually. With these it is better to sow the seeds as directed, and judge by the results.

To determine the quality of the seed, the normal germination capacity (as a percentage) needs to be known, as well as the length of time (in months or years) for which the seed remains viable, and the germination time in days or weeks. The test should be carried out at the optimum temperature for that species.

The test can be carried out in various ways. The following is a simple procedure: a plate or a seed tray without holes is lined with two or three layers of absorbent paper (kitchen paper), which is kept moist. A specific number of seeds is placed on to the paper. The seeds must be taken at random. On no account should you select the seeds for the test; they must be a representative sample. The more seeds are taken for the test, the more accurate the result calculated from it will be. Usually 10–100 seeds are taken. If the seeds are particularly fine, it may be necessary to measure by volume, and calculate the total number by multiplying up. Very large seeds (peas and beans) can be placed on pure sand if wished. When the seeds are in place, the container is covered with a sheet of glass or clear plastic. This creates a high level of humidity underneath it and prevents the seed from drying out. The test container is put in a warm place at 18–24⁰C (64–75⁰F) (in a room, for

A germination test can be carried out using a plate and absorbent paper

example). Everything must be kept very clean and hygienic throughout the test, because the conditions are not only right for the seeds to germinate but for fungus spores as well.

When the set period is up, the number of seeds which have germinated is counted. The number of germinated seeds is divided by the original total number of seeds in the test, and multiplied by 100 to give the percentage. This is the germination capacity of the seed sampled.

The set period is the time normally taken for seeds of that species to germinate. It is possible to allow a longer period before carrying out the count. Cucumbers usually take ten days, but the count could be done after fifteen days.

If the germination capacity of vegetable or herb seeds is below 20 to 25 percent, it is hardly worth sowing them unless those seeds are no longer available. There are many tree and flower seeds which cannot simply be put to a germination test, particularly those which need to pass through changes in temperature before they will germinate, and those which need to be left to lie for some time, or to be stratified.

Harvesting seeds

Is there any value in collecting your own seeds, if professionally produced ones are available cheaply? The professional seed growers are constantly endeavouring to produce seeds with the characteristics which the customer expects and wants, especially in the case of flower and vegetable cultivars. Consider, for example, seeds harvested from F1 hybrids. These do not produce plants with identical characteristics to those of the parent plant from which the seeds were collected. They will be divided into groups, each exhibiting some of the characteristics of either or both parents. F1 hybrids are produced by crossing two strains (parent plants). The first daughter generation is the F1 hybrid.

Various species can be propagated without any great problem from seeds collected yourself. The important point is to make sure that the seeds are fully mature when you collect them. Some do continue to ripen after they are gathered, but with many others, they only germinate satisfactorily if the seeds were already mature when harvested. Often, the seeds do not ripen all at the same time, and harvesting needs to be done several times. There are also plants which suddenly release ripe seeds. These include seeds which ripen in capsules, like the pansy or balsam. Choosing the right moment to harvest the seeds is important. Very rare or valuable seeds which need to ripen, but are then suddenly released, have to be protected.

*Seeds of the cycad (*Cycas revoluta*) have to be sown at 30–35⁰C (86–95⁰F), and may take months to germinate*

The seed heads are surrounded with a parchment or woven wrapping to catch the seeds. Seeds which ripen inside fleshy fruits like cucumbers or tomatoes need to be separated from the flesh of the fruit and cleaned before being dried. It is quite easy to separate out the individual seeds by hand if only a small quantity is required, and then rinse and dry them. Otherwise, the method is to cut out the inner flesh with the seeds, and allow it to ferment in water for a little while. It is important to make sure that the temperature does not become too high, as this could adversely affect the seeds' capacity to germinate. As soon as the flesh has liquefied, the fermentation process should be stopped. The liquid is poured off through a sieve with a suitably sized mesh. The seeds are then collected and rinsed clean. Finally, they are spread out to dry in a temperature between about 20 and 30⁰C (68 and 86⁰F). Seeds should only be put into bags, jars, boxes or other containers for storage once they are properly dry. Otherwise they could easily become mouldy, and this would quickly destroy the seeds.

Seeds of the loquat (Eriobotrya japonica) *are found inside a tart but very pleasant-tasting fruit*

A great many exotic fruits are on sale in the shops today. Many of these have seeds which can be sown. However, they do not always germinate very well, largely because the fruit is picked before it is fully ripe to help it to keep better, and make it easier to transport. Subsequent ripening of the seeds is not always enough.

It is also possible that some fruit may have been irradiated to make it keep better (though any irradiated food should be labelled). Irradiation will almost certainly destroy the capacity of the seed to germinate. (See page 94 for further tips on obtaining seed from exotic fruit.)

Conifer seeds ripen in cones, which usually open suddenly and fall to the ground. The seeds are obtained by a process of **seed extraction**. The cones are placed on pallets and subjected to considerable heat. After a quite short time the scales of the cone open and the seeds fall out but vigorous shaking can also help to release the seeds from the open cones.

Preparation of the seed

The seed obtained can be prepared in various ways which improve the quality. Thorough cleaning and removal of foreign material is the first important task. Then the seed can be sorted by size (**graded**), which makes mechanical sowing easier. **Pelleted** seed is another type which is easier to sow. Pelleted seed is also helpful to anyone growing plants like radishes or carrots on a small scale, because it is then unnecessary to thin them out. Another aid to sowing is to use seeds supplied sandwiched in between layers of cellulose or some other water-soluble substance, and arranged at a suitable distance from each other. The same principle applies to **seed bands**, which are simply laid in the prepared ground and watered in. The distance between the plants is then ideal.

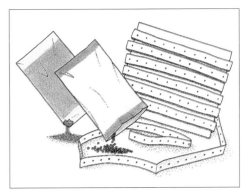

Presentation of seeds:
Left: normal seed; centre: pelleted seed; right: seed band

Seed storage

Seed can be stored for differing lengths of time. It is essential to realize that the embryo inside the seed is respiring. This uses up the reserves inside at a greater or lesser rate, and that in turn affects the length of time for which the seed remains viable. If the water content of the seed is lowered by slow drying, it can be stored for longer. The enzymes involved in respiration are activated less, the breakdown of carbohydrates is reduced, and the seed is able to be stored for a longer period.

Since the rate of respiration is also affected by temperature, the seeds should be kept cool, ideally just above 0^0C (32^0F). This explains the advice so often heard, "store in a cool, dry place."

Dormancy and promoting germination

Often, seeds do not germinate, even if the conditions for it appear to be ideal. Various causes may lie behind this failure to germinate. For some species of plant, the dormancy may even be vital to their survival.

In such cases the following causes may be present:

Freezing seed
Well-dried seed from a number of plants can be frozen at -20^{0}C (-4^{0}F), and kept for some time. It should be thawed slowly before sowing. This method does not however apply to a large number of tropical seeds.

● The seed coat is very hard, and water cannot enter, so that the activity of the embryo cannot be stimulated.
● The primordium is not yet fully developed, and needs more time (after-ripening).
● The substances inhibiting germination inside the seed or in the flesh of the fruit are not yet broken down.

More than one of these causes may be present together.

If plants are to be raised from seed, it is important to overcome dormancy. There are several methods. **Chipping** seeds with a very hard seed coat before sowing has proved very useful. This is a process in which the seeds are turned in a drum together with pieces of broken glass or cuttings of iron, for 15 to 45 minutes, depending on the seed coat. A possible treatment for some seeds is to rub them for a while with sandpaper.

Dormancy of eucalyptus seeds can be overcome with hot water

The dormancy of some seed species is overcome by pouring **hot** or **boiling** water over them. (Examples are species of robinia and eucalyptus.) Although it has been done, not infrequently, it would be better not to try treatment with concentrated sulphuric acid.

If development of the embryo is not yet fully complete, the seeds must be allowed to rest for whatever length of time is needed. Development is particularly marked if the seed lies in moist sand. Once germination has occurred, the seed must be sown straight away. It is not possible to re-dry it.

Seed treatment

Seeds are treated (dressed) to protect them from harmful organisms, particularly fungal infections. Harmful organisms are often found in unsterilized growing media, or on the surface of the seeds themselves. When packing firms treat seeds, the substances used must be declared on the seed packaging. It is possible to treat seeds which you have collected yourself, or which were acquired untreated. Only approved substances may be used. These are usually preparations in powder form, which are put into a container with the seeds and shaken vigorously. Excess is sieved off. The approved dressings only adhere to the seeds in tiny quantities. This is normally enough to protect them from diseases. The substances used for the dressing will have been broken down by the time harvest arrives.

Not every amateur gardener is prepared to use seed dressings, however. In this case the advice – even though it is more expensive – is to use a steam-sterilized growing medium, and to

How not to handle seed dressings: When treating seeds yourself, you should always wear protective gloves!

observe hygiene when working. Seeds can be bathed in a disinfectant solution (quinosol for example) before sowing. Covered plant trays should be aired from time to time. Coated seed, in which the coating around the seed contains natural and other protective substances, can also be a help. This layer dissolves after sowing, and gives protection to the emerging seedling.

Safety note
Suitable protective measures should be taken whenever working in ways which may involve coming into contact with chemical substances. Protective gloves should therefore be worn when treating seeds with dressing. Avoid breathing in the chemicals.

Stratification

Some seeds will only germinate reliably if they have been stratified first. Stratification means storing in layers. This storage imitates what would happen to the seeds out in the wild. The seeds of many temperate trees are stratified before sowing. The seeds of the species in question are mixed with moist quartz sand, filled into containers of a suitable size and stored in a cool place. Sphagnum moss peat can also be used if stratification is to be carried on for a long time, because it holds the moisture better. The medium should be sterile. Storage can be in a shady place in the open, or in a cool store under controlled conditions. The most beneficial temperature range is between 0 and 5°C (32 and 41°F), so stratification in a domestic refrigerator is usually not possible. The usual period of stratification is four to sixteen weeks, but

sometimes very much longer. At the end of the stratification process, the seeds begin to germinate. They must then be sown immediately. If they have to wait any length of time, the roots or shoots may break, or the root neck will grow noticeably crookedly.

Although it is a normal occurrence in nature, there is no particular benefit in the effect of frost on the layers of seeds. But it does no harm. Even the seeds of some subtropical plants germinate better for being stratified. Amongst these are the olive (*Olea*), jujube (*Ziziphus*), persimmon (*Diospyros kaki*) and other species of ebony (*Diospyros* sp.).

At the moment, a fashion seems to be spreading across Europe for the growing of ginseng (*Panax quinquefolius*, *Panax ginseng*), which is increasing in Germany, Austria, France and Switzerland. A number of agriculturalists and amateur gardeners have been intrigued by the massive profits which have already been achieved in parts of the USA and Canada, and especially North Korea, and are interested in trying their luck with the cultivation of this relatively still unknown, expensive medicinal plant. It is only possible to obtain ginseng by using stratified seed. If it is not stratified, the seed may take up to three years to germinate; sometimes it takes even longer, or fails altogether. The stratification of ginseng can take up to twelve months or more.

The length of time involved brings the risk of fungal infections. The seed will be ready for use in the following autumn. It must be sown straight away, or the seeds would germinate, become entangled and eventually spoil. The process can be delayed for a while by cool storage.

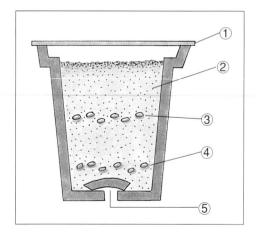

Stratifying seeds in a flower pot:
①*Protective cover* ②*Layers of sand*
③+④*Seeds* ⑤*Drainage hole with crock*

Sowing

Seed sowing in the open

If sowing is done in the open ground, the **seed-bed** should be prepared in advance. A particularly good type of soil is a humus-rich one with a natural crumbly texture, one that is not inclined to become waterlogged. Loam soils should be loosened and the texture improved as necessary so as to make them free-draining.

Seed should if possible be sown in **rows,** to make it easier to tend the young plants. The rows are set parallel to each other; the distance apart depends on the species of seed being sown. For most species, 25cm (10in) is the optimum distance, for radishes, a good 10cm (4in) is sufficient, and cabbage should be 50cm (20in) apart.

The depth of the drills also depends on the plants to be raised; 3cm (1in) is average. The seeds should not be sown too thickly, as the plants will in any case need to be thinned out when they come up. After the seeds have been sown, they are covered, and the surface lightly patted down with the back of the rake.

Using pelleted seeds makes it particularly easy to place them the desired distance apart. With these, it is especially important to moisten the soil and the seed coating thoroughly, or germination may be delayed.

Using seed bands, sowing is even easier. These are simply laid along the prepared drills. The first watering is given before the drills are covered, then they are filled in with soil and lightly patted down, and finally they are watered again. The distances apart will be correct from the outset, and there will be no more thinning out.

Some seeds are not sown in rows but in groups, with several seeds in each group. This is the normal method for various species of beans. Sowing in this way means that the group of seedlings will come up together, which is an advantage on heavy soils which tend to puddle. Seeds which need light to germinate, for example various species of conifer, are covered with a thin layer of quartz sand after sowing, and watered. If they have been stratified, they can be sown along with the sand which was used for the stratification process.

Sowing markers

When sowing seed (e.g. onions and carrots) which lies for a long period before germination, especially in low temperatures, quick-germinating species such as radish or lettuce should be mixed with it. The earlier-germinating seedlings will then mark out the rows and make cultivation easier.

Sowing in rows:
Carrots and onions grown together

Seed sowing in open ground:
①The soil of the seed bed is well loosened and levelled, and raked to a fine tilth

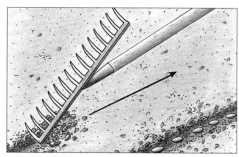

④Sowing in rows: the seeds are placed at approximately the required intervals along a drill made with a stick or the rake

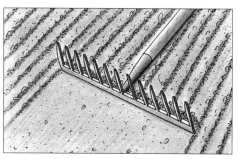

②The seed bed is smoothed carefully with the back of the rake

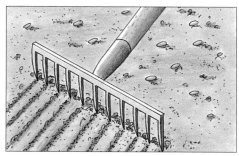

⑤Broadcasting: seeds which need darkness to germinate should be covered with soil

③Broadcast sowing is used for fine seed, or if the plants are to be grown continuously across a wide area

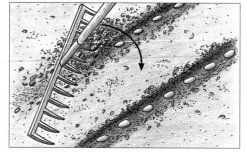

⑥Sowing in drills: seeds which need darkness to germinate should be covered with soil

Watering

The ground into which seed is sown, and in which the young plants will grow, should not be allowed to dry out. Seedlings and small plants can otherwise die off very easily. At the same time, ground which is too wet is also very detrimental. The seedlings are often quite fleshy, and, if the ground is too wet, they can then become rotten or fall victim to fungal infections once they are weakened; sprouting seeds suffocate for lack of oxygen, or rot. A constant, balanced amount of moisture should be provided at regular intervals.

If sowing is done into good seed compost (of the soil-containing type), problems should not arise, because the medium is free-draining, yet does not dry out easily. Seed compost is also stable in structure and does not tend to clog. The seeds should not be watered with too concentrated a jet, but with a very fine spray. Otherwise the seeds can be washed out of the soil, and the young plants damaged. This is especially true of seeds that need light to germinate, for example, some herbs (basil, camomile, thyme) as they are either left uncovered, or are covered only very lightly with soil. In this case, very great care should be taken when watering the ground.

It is only necessary to **fertilize** the sowing medium if it is entirely free of nutrients, and then it should only be done lightly. The fertilizer already present in the medium or, in the case of home mixtures, the organic matter, is sufficient until the seedlings are transplanted (pricked out).

Sowing in seed trays and containers

More tender seeds are not sown straight into open ground in spring, but first into **seed trays**, which are covered with a

sheet of glass. This protects them from the less clement spring weather, and from any frosts. The seed can germinate and begin to grow in protected conditions, but the young plants will need to be hardened off gradually when the weather is suitable. To do this, the cover is removed for a gradually increasing period of time.

Sowing in trays is not normally done in rows, as it is in open ground, but **broadcast**. Since the seedlings will have to be transplanted in any case (see page 31), there is no extra work involved in doing this; sowing in this way also enables maximum use to be made of the space in the seed tray, which is usually limited. The seed compost used in the tray must be sterilized (by steam), because the broadcast method of sowing means that it is not really possible to cultivate around the seedlings.

Some subjects are sown in **pots,** because this makes the eventual transplanting easier, and the roots of the seedlings are not damaged. The main plants in this category are pumpkins, melons and cucumbers, especially the valuable F1 hybrids. The fig-leafed gourd (*Cucurbita ficifolia*), which is popular as a rootstock for cucumbers, is also raised in pots. This is necessary so that the grafting can be carried out without difficulty. Other species of vegetable too could be initially raised in pots; this would make transplanting considerably easier, and promote establishment and subsequent growth. This does however require a considerably larger propagating area, and this is why only selected species are grown in this way.

①

③ a

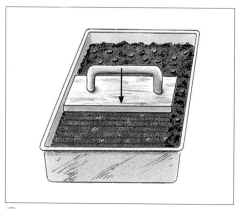

②

③ b

Sowing in seed trays:
① *The seed compost is levelled and smoothed, and the seed scattered evenly*
② *A wooden block is used to press it down*
③ *(a) Soil is sieved over seeds which need darkness to germinate* ③ *(b) Seeds which germinate in the light remain uncovered; both are moistened thoroughly*
④ *A cover to prevent drying out is important. After the seeds have germinated, the cover is lifted more and more to provide ventilation. Finally it is removed altogether*

④

Thinning and transplanting

When seeds are sown into open ground, the plants are **thinned** as soon as they are 5–10cm (1½–3in) high. The weakest are pulled out, to leave the optimum distance between the remaining plants.

Plants which have been raised in seed trays must be hardened off before they are planted out. To do this, the glass cover is removed more often and for longer each day, until the plants are no longer covered. The strongest plants are transplanted or **pricked out**, either directly into open ground, or initially into pots 9 or 10cm (3½ or 4in) in diameter, to give the plants the best chance of developing further without competition. When all likely risk of frost is over, from early summer, they are planted into humus-rich and nutrient-rich soil in open ground. Exotic plants sown as seeds directly into pots should continue to be grown in the same pots until these become too small for the plant. Then they should be potted on into a slightly larger container. It is important to make sure that the root ball remains intact. Pruning back the root is not necessary for the young plants. Standard soil, prepared by fertilizing, is recommended as the growing medium. The best time for potting on is spring, just before new growth begins.

①

②

Pricking out:
The first foliage leaves have now been formed above the two seed leaves
① Take gentle hold of the seedlings between thumb and forefinger, and lever out with a dibber
② Insert with the aid of the dibber. Allow sufficient room for the root, and take care not to bend it sharply
③ Ease compost gently but thoroughly around the seedling

③

Propagation by vegetative means

Vegetative propagation – in contrast to propagation by seed – is asexual increase. This type of increase also occurs in nature; many plants multiply asexually as well. A typical example is the formation of offshoots by many species of the banana family (Musaceae). Particular species of agave (for example *Agave americana*) also often form a large number of offshoots. Bromeliads (Bromeliaceae) form offsets, and the spider plant (*Chlorophytum*) runners. Plants which form bulbs and corms increase by means of bulblets and cormlets. All these produce new plants which are identical to the parent plant.

Vegetative propagation using portions of plants is by far the most widespread means used in horticulture. It has in many ways considerable advantages over propagation by seed. Vegetative means of propagation are used for the following reasons:

● The plants raised in this way are identical to the parent plant. The characteristics and appearance of the new plants will be known in advance. Particular selections can be propagated and specific traits retained in the process.

● Dioecious plants (that is, plants that have male and female flowers on separate specimens) can be propagated to give a plant of the required sex, for example kiwi (*Actinidia*), pistachio (*Pistacia*), sea buckthorn (*Hippophae rhamnoides*).

● It may take many years of cultivation for some plants to assume their mature forms. Vegetative propagation can produce these features even in quite young plants, for example eucalyptus (*Eucalyptus*).

● Depending on the method chosen, large plants can be produced much more quickly by vegetative means of propagation such as air layering (see page 46) rather than by seed.

● Vegetative propagation in horticulture is usually less expensive, because it can be done quickly, at lower cost, and with predictable results.

Cuttings

Cuttings are portions of the parent plant which have been removed from it; they are then made to root, producing a new, independent plant.

The term "cutting" is usually understood to mean a stem or section of stem with leaves. However, buds,

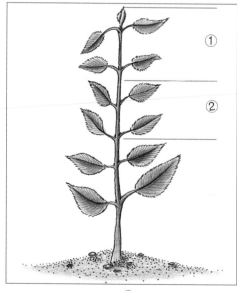

Types of cutting: ① Tip cutting ② Softwood stem cutting

32

individual leaves, or portions of leaf also come into the category of cuttings. In general, deciduous plants – which include many temperate trees – are propagated by cuttings which are not fully ripe, that is, not hardwood cuttings. Cuttings from evergreen plants often root better if fully ripe cuttings are used. Meristem culture (see page 98) is also a type of vegetative increase, although the growth characteristics of plants propagated in this way can often resemble those of a plant obtained from propagation by seed.

How to take cuttings

Portions of any plant can in principle be removed and used to create new plants. Cuttings of some plants are easier to root than others, which may be difficult or impossible to root, but, apart from this, it is mainly the parent plant that determines success. The age of the parent plant of various species has proved to have a decisive effect on the rooting of cuttings. Using a **younger** parent plant is preferable for many species. In horticultural practice, cuttings are even taken from rooted cuttings which have begun to produce shoots, in order to achieve a high degree of rooting success.

The **health** of the parent plant is important. It should be well nourished and making good growth, with its growth constantly renewed by cutting back. Cuttings taken from older plants often show a definite decline in their capacity to root. If no other plant is available, it is possible to achieve a noticeable improvement in the ability of the cutting to root by fertilizing the parent plant in a deliberate and specific way before the cutting is taken.

The cut is made through the stem , just below a leaf bud or eye (a node),

because this is where there is a high concentration of nutrients. This will have a positive effect on the formation of adventitious roots (that is, roots formed on the shoot).

Although it is usually thought necessary to use a very sharp knife for cuttings, this need not be the case. Taking cuttings can be done more simply and much more quickly with **secateurs**. Another common, traditional technique is to cut back the leaves to reduce moisture loss, but again, this is not necessarily always to be recommended. This is because the leaves contain compounds manufactured by the plant and stored there. They may be used by the plant in its later growth, but cutting back the leaves will remove them.

When to take cuttings

There are a number of factors which determine the best time, so that a set time for the various species cannot be given. As a general rule, species which root easily are less dependent on timing, and some even allow cuttings to be taken successfully all year round. Species which are difficult to root do depend for success on choosing as closely as possible the precise time which is best for taking the cutting.

Research findings show clearly that the best time is when the concentration of starch and the proportion of carbohydrates in the cutting are at their highest. These data are very difficult to determine, especially of course for the amateur gardener.

The best advice is therefore to insert fresh cuttings without delay, once they have been taken.

Propagating plants by cuttings:
① Sever the cutting from the stock plant

③First dip the cutting in rooting powder;
then insert it into the compost

Choosing the best time of day

Cuttings should be taken in the early morning or afternoon. In the morning, the plant is particularly well supplied with water (turgid), which has a positive effect on early survival. In the afternoon, the carbohydrate concentration is higher. This has a positive effect on rooting. If it is possible to insert the cuttings immediately, the afternoon is to be recommended.

② Remove the lower leaves

How to insert cuttings
It is usual to insert at least one bud or pair of buds. Occasionally, wounding the bark or rind is also recommended as this is said to encourage better and quicker rooting.

More than one bud can be inserted, especially if the gaps between buds (**internodes**) are particularly short. The cut can be sprinkled with powdered charcoal; this has a disinfectant effect, and helps prevent decay. At least one bud should be above soil level.

Cuttings are inserted into cutting compost. This compost consists of sand and/or peat; usually a 1:1 mixture of both is used. The compost should not contain fertilizer or nutrients (see page 17). The addition of perlite or vermiculite, or similar substances to increase the water-holding capacity of the compost is recommended. It is important to firm in the cuttings by pressing down the compost lightly around them. This makes sure that they are fully in contact with the soil.

Cuttings need some bottom heat and a moist atmosphere. It is advisable for this reason to spray them frequently, or to put clear plastic or glass over them. A high level of humidity can then build up underneath. Cuttings of non-native species in particular should therefore be grown in a propagator or greenhouse. Small indoor propagators are also very suitable for this, especially those which have adjustable heating in the base.

Rooting cuttings in water

Cuttings from some plants root very easily, and will even do so without difficulty in water. For this, cuttings 15–25cm (6–10in) long are taken from the plants in question. The leaves are removed from the lower part of the stem, and the cuttings put into a glass of clear water without plant nutrients. After a while, roots form, and the young plants can be potted up and either hardened off or planted out.

The delicate, glassy roots are very vulnerable, so it is wise not to wait too long before planting out. Otherwise it is easy to break off roots which have already become quite long.

Plants which can be propagated particularly well by rooting cuttings in water are: Japanese quince

Rooting cuttings in water

(*Chaenomeles*), Cyperus alternifolius, dieffenbachia (*Dieffenbachia*), Epipremnum aureum, forsythia (*Forsythia*), ivy (*Hedera*), busy lizzie (*Impatiens*), privet (*Ligustrum*), oleander (*Nerium oleander*), African violet (*Saintpaulia*), willow (*Salix* sp.), coleus (*Solenostemon, Plectranthus*), African hemp (*Sparmannia*), tradescantia (*Tradescantia*).

How a cutting roots

In the growth region of their stems, plants usually have root beginnings. After the cutting has been inserted, the cell sap which is oozing out at the site of the cut or wound forms a fatty substance which seals the wound; this may develop to varying degrees. Wound tissue is formed in the **cambium**, the growing area of the stem. This tissue is called **callus**. In a suitable medium and under favourable conditions, new roots grow through or out of the callus; these are called **adventitious roots**. The thickness of the callus formation depends both on the plant species and on various external factors. A very thick layer of callus is not usually beneficial, because it can hinder root formation considerably; the new roots are normally formed

35

underneath the callus layer. The cutting is however able to draw water and nutrients through the callus, and so ensure its survival. Freshly inserted cuttings which have only reached the stage of producing fatty substance and sealing the wound have very little ability to take in water. The formation of callus is the first successful step towards rooting and ultimately becoming established.

Hardwood cuttings

One special type of cutting is the hardwood cutting. These are usually one-year-old stems, which have ripened and become woody, and have lost their leaves. They are normally taken from deciduous subjects in late autumn, before the first severe frosts, and planted in spring. They are stored in the meantime in cool conditions with high humidity,

Adventitious roots

When cuttings make strong growth without developing roots, they can be stimulated to form adventitious roots by artificially wounding the callus tissue.

for example in a cold store. In the absence of cold store facilities, hardwood cuttings can be stored by burying them in a shady place in the garden. A cool shed or garage may also be a suitable place. The cuttings should be stored in moist sand and wrapped in plastic film. Cellars, where available, are often too warm (5–10^{0}C; 41–50^{0}F). If the cuttings

Propagating by hardwood cuttings:
① Hardwood cuttings 20cm (7in) long are taken from the stock plant in winter

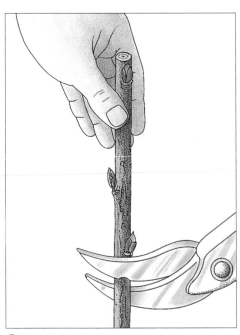

② The individual lengths are prepared by cutting through near a bud

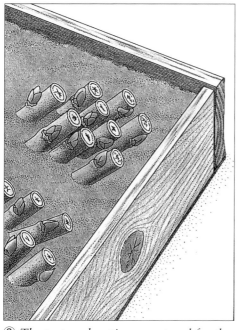

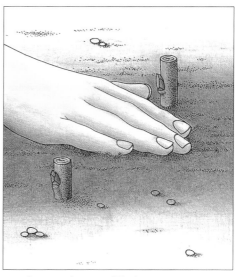

③ *The prepared cuttings are stored for the winter with the top bud still visible*

⑤ *The trenches are filled in*

④ *In spring, they are set at 10–15cm (4–6in) intervals upright in trenches. The top bud must be above the soil*

⑥ *When the cuttings have rooted, they are taken out and replanted where they are to grow*

are taken in frosty conditions, they should on no account be brought into a warm room. It is often advisable to take preventive measures against attack by pests and fungus, as well as preventing the cuttings from being eaten by rodents.

The cuttings are trimmed to at least two buds, though more usually five to six buds. This gives them a length of about 15–25cm (6–10in). The wood from the middle of the stem makes the best cuttings, because the buds are particularly well developed along that part of its length.

Secateurs are quite adequate for the job of preparing the cuttings. It is customary to bundle the hardwoods together, and cut the whole bundle to the desired size with a band-saw. However, for species which are more difficult to root, the cut should be made immediately underneath a bud. The greater quantity of reserves stored there facilitate rooting. The upper end of the prepared cutting should have the cut sealed with grafting wax to prevent drying out and infection caused by pests.

This treatment also makes it easy to distinguish one end of the cutting from the other after it has been stored for some time. It is essential to plant it the right way up.

Propagation by hardwood cuttings has a number of advantages, even if the same plant can also be propagated by other types of cutting. It is possible to take the hardwood cuttings in winter, a time when there is usually less to do in the garden. Hardwoods are more robust, and finally, it is possible to raise plants of many species in the open without great fuss and expense or special equipment.

The usual time for planting is mid to late spring, when no more hard frosts are to be expected. Planting in pots in a protected spot can be done considerably earlier.

Fruit rootstocks for grafting are propagated to a very great extent by hardwood cuttings, as are many ornamental trees and shrubs, and woody plants from warmer regions. These

Plants which can be propagated by hardwood cuttings

Bastard indigo (*Amorpha fruticosa*), *Ampelopsis*, *Callicarpa*, Japanese quince (*Chaenomeles*), clematis, traveller's joy (*Clematis*), dogwood, cornel (*Cornus*), smoke tree (*Cotinus*), cotoneaster (*Cotoneaster*), quince (*Cydonia oblonga*), deutzia (*Deutzia*), oleaster (*Elaeagnus*), true fig (*Ficus carica*), forsythia (*Forsythia*), sea buckthorn (*Hippophae rhamnoides*), crape myrtle (*Lagerstroemia indica*), honeysuckle (*Lonicera*), apple rootstocks (*Malus*), mulberry (*Morus*), Virginia creeper (*Parthenocissus*), philadelphus, mock orange (*Philadelphus*), plane (*Platanus*), knotweed, polygonum (*Polygonum*), poplars (*Populus* sp.), potentilla, cinquefoil (*Potentilla*), plum rootstocks (*Prunus*), currants, gooseberries (*Ribes* sp.), roses (*Rosa multiflora, Rosa nitida, Rosa rugosa*), willow (*Salix*), elderberries (*Sambucus* sp.), spiraea (*Spiraea*), snowberry (*Symphoricarpos*), lilac (*Syringa*), tamarisk (*Tamarix*), elm (*Ulmus*), highbush blueberry (*Vaccinium corymbosum*), Guelder rose (*Viburnum opulus*), grape vine (*Vitis*), weigela (*Weigela*).

should be planted in a warmed propagating medium, however (14–18°C; 57–64°F).

Leaf cuttings

Some plants can be propagated from their leaves. This method is especially suitable when there is only a little propagating material available. For example, this simple method of propagation is used for camellias (*Camellia* sp.), especially the tea plant (*Camellia sinensis*). A leaf which is just mature (that is, not an older one from the bottom of the plant, and not a young one which still has its glassy look) is removed together with a bud, and planted. If a great deal of callus is formed, but no adventitious roots appear, the best remedy is to wound the mass of callus tissue artificially. This often succeeds in bringing about root formation.

The African violet (*Saintpaulia*) is also easy to propagate from leaf cuttings. A single leaf is removed from the parent plant, and the leaf stem shortened to 1–2cm (¹/₂–³/₄in). The leaf is laid on the moist propagating medium, covering the stem lightly with soil. The soil is slightly warmed, to 22–24°C (72–75°F), and the air kept humid. Rooting will then take two to three months. Small plants will be formed on the stems, and these will soon need to be transplanted.

There are some species of begonia which are just as easy to propagate. The method involves cutting off a mature leaf of *Begonia rex* hybrids, and wounding the veins on the underside with a sharp knife. This leaf is then laid on to moist propagating medium and pressed down, or held down with small

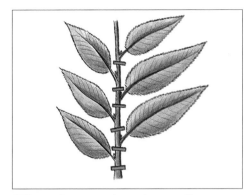

Leaf cuttings from the camellia: ①*The possible points for cutting are marked in red*

②*The cuttings are inserted. Clear plastic film keeps the right atmosphere inside*

③*Rooted cutting with new shoot*

39

Propagation by vegetative means

Leaf cuttings from African violet:
①The leaves inserted in the seed tray

②Small new plants form at the stems

③The small plants are potted

Leaf cuttings from begonias:
①Cuts are made into the large veins on the underside

weights to ensure that it stays in contact with the soil. In a heated (indoor) propagator, or other suitable propagation container, at about 24⁰C (75⁰F), the leaf will produce roots in several places after a few weeks. Then is the time to cut up the leaf into separate portions and pot the individual plants.

Begonia masoniana can be propagated from pieces of leaf a few centimetres (an inch or more) across. The leaf veins must be cut or wounded. These leaf cuttings are inserted vertically into the propagating medium, firmed in and given the same type of treatment as the *Begonia rex* ones, in high humidity. Once they have become established and produced shoots, they are transplanted and slowly hardened off.

Hybrids of the Cape primrose (*Streptocarpus*) are propagated from half leaves with the middle rib cut away.

The half leaves which result are pressed into the propagating medium with the cut side downwards.

To propagate *Sansevieria*, the long leaves are cut into pieces about 5cm (2in) long; these pieces are then inserted into propagating medium. Under the conditions mentioned above, they root

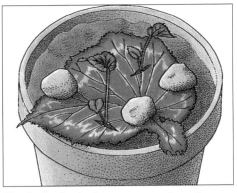

② *The leaf is laid on a peat and sand mixture, underside down and weighted with stones. Later, when the young plants have rooted, they are separated*

Plants which can be propagated by leaf cuttings

Begonias (*Begonia rex* hybrids, *Begonia masoniana*, *Begonia boweri*), tea plant (*Camellia sinensis*), camellias (*Camellia* sp.), crown of thorns (*Euphorbia milii*), opuntia, prickly pear (*Opuntia*), Easter cactus (*Rhipsalidopsis*), African violet (*Saintpaulia*), green varieties of *Sansevieria*, *Selaginella*, Cape primrose (*Streptocarpus*).

Sansevieria *leaf cuttings with new shoot*

without difficulty. It is important to make sure that it is the base of the cutting which is inserted into the growing medium.

Finally, many plants of the cactus type, opuntia and phyllocactus, can be propagated easily by leaf cuttings. Individual segments broken off the parent plant can be inserted as cuttings into propagating medium.

Mound layering

The propagation of plants by mound layering is a simple and economically efficient method of raising trees and shrubs true to type. The method, which is also called **stool layering**, is mainly used by tree nurseries, but it can equally well be used by any plant-lover.

The stock plant (or parent plant) is cut back hard in the autumn, which encourages the development of buds at the base. The shoots which arise in the following year are earthed up repeatedly until mid-summer, resulting eventually in a mound of earth 20–30cm (7–12in) high. The bases of the shoots therefore remain soft, and are stimulated to produce adventitious roots. When the shoots are fully mature in late autumn – all leaves must by then have fallen – the mound is removed. The rooted shoots are exposed, and are then cut off at the base and planted out either immediately or in the following spring. The stock plant is covered up again with humus-rich soil. The main plants propagated in this way are apple, quince, cherry rootstocks (F12/1), the buffalo currant (*Ribes aureum*), gooseberry (*Ribes uva-crispa*) and hydrangeas (*Hydrangea*).

41

Propagation by suckers

① + ② *The stock plant is severely cut back, and the young shoots are earthed up until late mid-summer*

A particularly easy method of producing new plants identical to the parent is by obtaining suckers. A number of woody plants, usually older ones, have a suckering habit. When the suckers have matured in late autumn, they are separated from the parent plant. The usual method with many varieties is to sever them using a sharp spade. Another method which is sometimes advisable is to expose the roots and separate the suckers with secateurs.

Many species of fruit and ornamental tree tend to produce suckers, for example plum (*Prunus domestica*), cherry plum (*Prunus cerasifera*), sloe (*Prunus spinosa*), quince (*Cydonia oblonga*), Japanese quince (*Chaenomeles* sp.), apple tree (*Malus*), sea buckthorn (*Hippophae rhamnoides*), chokeberry (*Aronia* sp.), stag-horn sumach (*Rhus* sp.) and others.

If fruit trees are to be propagated true to variety, they must be growing on

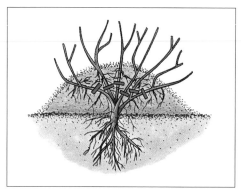

③ *After leaf fall in late autumn, the earth is removed. The rooted stems are severed right at the base*

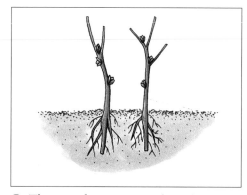

④ *The rooted stems are replanted either immediately or in the following spring*

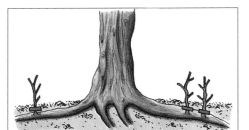

Suckers on a fruit tree, showing best position for cut

their own roots and not grafted on a rootstock. Otherwise, the suckers will be shoots formed by the rootstock and not shoots of the particular variety.

Similarly, **stolons** are formed by the strawberry (*Fragaria*), the spider plant (*Chlorophytum comosum*) and the sword fern (*Nephrolepsis exaltata*).

Propagation by division

Woody plants are seldom propagated by division today, because the benefits are not very great. The method can however produce plants which become strong specimens particularly quickly.

Plants which are especially suited to this method of propagation are berberis (*Berberis buxifolia*), creeping dogwood (*Cornus canadensis*) and – still used today – box (*Buxus sempervirens*). The important point in this type of propagation is the ability of the plant to form new shoots from its rootstock.

This method is particularly common for propagating herbaceous perennials, as well as orchids and plants which form bulbs or corms (see page 88).

Tropical and subtropical plants (*Strelitzia, Acca*) and grasses can also be propagated by division.

Most herbaceous perennials can be propagated by division. The root ball is separated into two or more pieces, each with some leaves or growth buds

Dividing houseplants

Many houseplants can also be propagated by division. For plants whose rootstock is particularly strong-growing, dividing is the method of choice, so as to keep the plant's growth under control. An example is the aspidistra (*Aspidistra eliator*). Many ferns can also be propagated by division.

Layering

Simple (ordinary) layering and continuous layering are similar to the mound layering method described above (page 41). Both simple and continuous layering are reliable methods of propagating plants, even those which are difficult to root. Both require a clear area of 1m (3ft) around the stock plant, with no other plants growing in it. This area is needed for propagation. In **simple layering**, a stem of the parent or stock plant is induced to form roots. This is the procedure: in spring, the stock plant is cut back strongly (that is, down to the ground) before it comes into new growth. During the year, the plant forms new shoots. Early in the spring of the following year, these are brought down into the soil in a curving line around the stock plant and held in place in the ground; the best way is with a layering pin or galvanized wire peg, the shape of an old-fashioned hairpin. The outer end of the shoot is turned upwards. If the stem is very hard or brittle, or if the plant is a species which is difficult to root, the stem should be twisted at the point where it turns at the sharpest angle and pinned down into the ground. This creates lengthwise splits in the stem which make it less likely to snap off, and also stimulates root growth at this point. For most species it causes no harm even if the stems crack open a little at the point where they are buried, as long as they do not break off completely. This would separate them from the stock plant and prevent them from being supplied with water and nutrients during

Simple layering:
①A shoot is bent down to the ground. The middle section has its leaves removed

②The bend of the stem is inserted into a hole about 6cm (2¹/₂in) deep, pegged in place, and covered

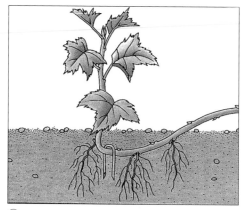

③Once rooting has occurred, it is separated and replanted

the rooting period, which can take up to three years. Although layering involves time and effort, it continues to be used for some plants. Blackberries (*Rubus fruticosus*) root particularly well by this method, as does the Guelder rose (*Viburnum opulus*). The various species and varieties of rhododendron (*Rhododendron* sp.) normally take two years to root; magnolias (*Magnolia*) and witch hazel (*Hamamelis*) take as long as three years.

Woody plants such as species of viburnum (*Viburnum* sp.), the smoke tree (*Cotinus coggygria*), and lilac (*Syringa vulgaris*) are propagated using a variant of this method. It is not the year-old, woody stems which are layered, but the current year's growth, as soon as it has reached a length of about 30cm (12in). The stems are still soft and easy to bend. The tips of the shoots must protrude from the ground, and if necessary damp, humus-containing soil must be piled up over the layers a second time.

Constriction with wire can be used to force rooting to occur, both in this type of layering and in simple layering, and also mound layering. A piece of strong, flexible wire (such as copper wire) is wrapped around the stem at the base. As the stem grows and becomes thicker, the wire causes a blockage in the flow of sap which is the plant's transport system, damming up the photosynthate and nutrients, and producing root formation.

Continuous layering enables many more clones to be produced than the simple layering procedure described above. Preparation is done in the same way. After the stock plant has been cut back to ground level in winter, it produces shoots vigorously, and forms long stems, which are allowed to grow for a year. In the spring following that

Continuous layering: The young plants are separated from the stock plant once they have rooted

year's growth, these stems are laid horizontally on the ground and fastened down. Strong layering pins are suitable for this. A number of shoots develop from the growth buds (**nodes**) along these horizontal stems. During the year – depending on the amount of growth made by the shoots – moist, humus-containing soil is heaped up around each shoot, keeping the tip free above the soil. This may need to be done three or four times. The soil level will then have risen 20–30cm (7–12in).

The layers are separated in late autumn or winter, when they have matured fully. The entire stem is dug up and the individual layers cut off. During the most recent growing season, the stock plant will have developed new stems, which can be layered in their turn. The separated layers are planted out immediately, or potted. But they are also often kept in a cold store until they can be planted in spring, or loosely planted. This method of layering can be used as a very easy way of propagating species and varieties of hazel (*Corylus* sp.).

Air layering

Air layering, sometimes called marcotting or Chinese layering, is very similar to simple layering. The term really says everything about the procedure: it is done above ground, in the air, and the plant is induced to form rooted shoots, or layers.

Air layering is not a new technique; there is evidence documenting its use for well over 1,000 years. Although it involves some effort and does not produce large numbers of plants, it has not slipped into oblivion even today. In southern countries, air layering is used to propagate from woody plants which do not otherwise root easily, or when large, strong specimens of the plant are required quickly. It is also a way of reducing the size of houseplants which have outgrown the space they are in, such as *Fatsia japonica*. In Israel and South Africa, the lychee (*Litchi chinensis*) is propagated in this way. It is quite possible for freshly rooted plants to blossom and bear fruit in the year they are planted out.

To carry out air layering, a stout stem is chosen; it is also possible to use one several years old. A slanting upward cut is made into the stem, reaching barely halfway through it. The cut is held open with a small stone or other object. A sleeve of plastic film is then tied around the stem, below the cut. This creates a pocket which is filled ideally with moist sphagnum moss; otherwise with peat or

Air layering:
① *A stout stem is cut halfway through. A small stone is wedged into the cut to keep it open*

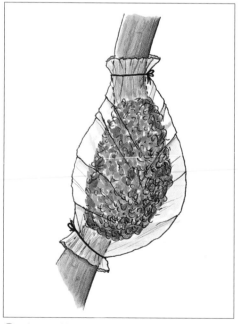

② *Plastic film is tied around the stem below the cut. It is filled with moist sphagnum moss or peat, and then tied in place above the cut*

soil. The sleeve is then tied around the stem above the cut in the same way. The medium used must not be allowed to dry out, so must periodically be remoistened. To do this, the top of the sleeve is untied, unless an opening for watering has been incorporated. For some species, rooting occurs in a matter of weeks; others take a year or more. Air layering should be carried out in spring. Then the plant has plenty of time to form roots before the autumn months. If the plant is being grown all year round in a heated room, perhaps a conservatory or sun room, air layering can be done at almost any time of the year.

After rooting has taken place, the new plant is cut off below the rooting site. It may be necessary to check whether rooting has occurred, if you are unsure, before severing the layered stem.

Plants used for air layering already have quite thick stems, so the cut on the stock plant should be painted with pruning compound or grafting wax, if only for the sake of the plant's appearance. The sleeve is removed, and the new plant potted into a container of sufficient size for the plant to grow on. It is advisable to care for the new plant very carefully at first, according to the needs of that particular species, as transplanting is a considerable shock for it. Another possibility is to cut open a plastic flower pot and use that to form the sleeve containing the medium used, instead of the plastic film. Then it may be possible for the new plant to remain in the pot for a while, after severing.

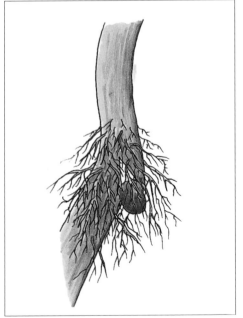

③ Roots form at the site of the cut. The time taken depends on the species. The new plant is separated by severing below the rooted area

Plants which can be propagated by air layering

Species of begonia (*Begonia* sp.), citrus species (*Citrus* sp.) tree ivy (x *Fatshedera lizei*), *Fatsia japonica*, true fig (*Ficus carica*), species of fig (*Ficus* sp.), lychee (*Litchi chinensis*), magnolia (*Magnolia*), Swiss cheese plant (*Monstera deliciosa*), olive (*Olea europaea*), *Philodendron*, rhododendron, azalea (*Rhododendron*), *Schefflera*.

Root cuttings

Some plants can be propagated vegetatively from root cuttings (pieces of root). The plants to be propagated are dug up in late autumn or winter, once they are fully mature. A few roots of at least pencil thickness are taken off, and the stock plant replanted, or heeled in and kept in a cool place until the spring; it depends on the weather. The roots obtained are cleaned and cut into pieces 4–8cm (1¹/₂–3in) long.

It is important to plant them the right way up, because even root cuttings produce shoots from their upper end and adventitious roots from the lower end. Cutting them differently at each end (for example, using a diagonal cut at the lower end) makes it possible to tell the ends apart at a later date. The prepared pieces of root are laid in a box of moist peat or sand and peat mixture, covered with the same mixture, and stored in a cool place until they are to be planted. The root cuttings are planted in boxes covered with a lid of glass, or in pots. Moist propagating compost made up of peat and sand is suitable. Thick root cuttings are inserted vertically or slanting. It can be helpful to leave the very top of the root cutting uncovered by the compost. This stimulates it to form a growth bud at the top and adventitious roots towards the bottom. Thinner root cuttings may be planted horizontally, and covered with a good 1 cm (¹/₂ in) of compost. The compost must not be allowed to dry out at any time.

Some species tend to produce a growth bud and shoot quickly, but without forming enough of the adventitious roots necessary for the new plant to grow. In this case a good solution is to earth up the shoot. This can cause the shoot to form roots more quickly than the root cutting itself.

The plants obtained by this method are hardened off gradually during the spring months, and then planted out in the open or wherever they are intended to grow.

Plantlet-bearing species

It is quite common for plants to propagate vegetatively of themselves, and

---

Plants which can be propagated by root cuttings

Aralia (*Aralia*), horseradish (*Armoracia rusticana*), Japanese quince (*Chaenomeles*), globe thistle (*Echinops*), Siberian ginseng (*Eleutherococcus* sp., syn. *Acanthopanax*), euonymus (*Euonymus*), meadowsweet (*Filipendula*), geranium (*Geranium*).

---

so pass on all the characteristics of the one parent plant to the offspring. A typical example of this happens when plants produce on their leaves, stems or elsewhere entire new plantlets with their own independent root system. When the plantlets have matured and grown sufficiently, they separate themselves from the parent plant and fall off. If they land in suitable soil, they take root and continue to grow.

The poet Johann Wolfgang von Goethe made a study of these interesting

*Piggyback plant (*Tolmeia menziesii*)*

Kalanchoe diagremontiana

plants. He was fascinated by this unusual means of plant propagation, and produced a description of the species *Kalanchoe diagremontiana*. Their leaves are fleshy and toothed, and in each notch is a bud which grows unrecognizably at first, but eventually each bud produces a tiny new plantlet with roots. In this way, each leaf can produce a large number of new plants. As soon as they have recognizable roots, the plantlets can very easily be removed by hand and planted in propagating compost. The Germans even call this *Kalanchoe* 'Goethe tree,' because he devoted so much attention to it. There are various other plants which also give rise to plantlets. The following is a list of them.

Plantlet-bearing species

Allium cepa var. *viviparium*, fern species (e.g. *Asplenium bulbiferum*), *Kalanchoe crenata*, the Mexican hat plant (*Kalanchoe diagremontiana*), *Kalanchoe laxiflora*, *Kalanchoe pinnata* (syn. *Bryophyllum calcinium*), *Kalanchoe prolifera*, *Kalanchoe tubiflora*, the piggyback plant (*Tolmeia menziesii*).

Grafting

Why should plants be grafted when so many other simpler methods of vegetative propagation exist?

Propagation by cuttings, layering or air layering is not always a possibility. The plant must be able to develop roots if it is to be propagated by cuttings, and the same applies to layering and air layering. But even if rooting does succeed, there may be reasons for preferring the method of grafting, because it is the only way of achieving the desired result.

Grafting is a type of vegetative propagation; the technical term for it is **xeno-vegetative (foreign-vegetative) propagation**. The process creates one individual plant from two, and the one which is grafted on usually predominates. But (with some exceptions) both plants used are necessary for the survival of the new plant.

An interaction is set up between the **scion**, the shoot grafted in, and the **rootstock**, the rooted section on which it is grafted. The rootstock supplies the plant with water and nutrients, and the scion provides the photosynthates, the compounds made by the leaves. So each part of the newly created plant affects the other. The rootstock usually has the most significant effect on the growth of the plant – it is often the main reason for choosing to graft at all. Meanwhile the scion retains its essential charac-teristics. A grafted apple tree is a typical example. The rootstock determines how vigorous growth is, when fruiting will begin, and what type of soil the tree will tolerate. The variety of the scion will be the variety produced.

A Cox's Orange Pippin will not turn into a Gravensteiner, or anything else; it will remain a Cox's Orange Pippin. But the rootstock can have a degree of influence on the size of apple produced, for example, or the colouring. The same sort of thing applies to other fruit species.

Grafting is done for the following reasons:

● To propagate plants which root poorly if at all, e.g. propagation of cacti which lack chlorophyll, such as yellow and red cacti.

● To propagate hybrids that cannot be increased by seed, e.g. propagation of the intergeneric cross between pear and quince, *Pyronia* (*Pyrus* sp. x *Cydonia oblonga*).

● To propagate "sports," shoots which display mutations, e.g. commercial nurserymen would use this method to propogate the apple variety 'Royal Gala' or 'Tenroy,' which is a coloured sport of 'Gala'.

● To raise plants more quickly, or bring forward the harvest date, e.g. certain varieties of plum and peach.

● To produce plants that are able to tolerate particular types of soil, e.g. *Citrus* on Japanese bitter orange (*Poncirus trifoliata*), which enables the propagated plants to do well on slightly acid soil.

● To produce plants with a particular habit of growth, e.g. dwarfed fruit trees for commercial growing (because harvesting can be done more cheaply, by machine), and weeping forms of shrubs and trees such as roses or weeping willows.

- To produce the greatest possible number of identical trees from a small amount of material, e.g. budding new cultivars.
- To preserve genetic material from plants liable to be lost, e.g. grafting from an ageing or partly dead tree on a young rootstock.
- To be able to grow several varieties of fruit on one tree, e.g. several apple cultivars on a "family tree" (they fertilize each other).
- To be able to grow vigorous plants in a tub, e.g. pears grafted on dwarfing quince rootstocks.
- To protect plants from diseases or pests, e.g. passion-fruit (*Passiflora edulis*) grafted on resistant rootstocks.

Rootstocks

Plants cannot be grafted on absolutely any rootstock because they must be sufficiently closely related. Good results can normally be expected if varieties of the same species are grafted one on the other. But combinations of plants of different genera can form a permanent, close union.

Scions

For many plants, one-year-old branches are best for grafting. They should be straight and unbranched, and should have leaf buds, not flower buds. It is better to choose those that have short internodes (the length of stem between individual buds). Common opinion says that suckers from fruit trees cannot be taken for grafting, but this is not quite

so. What must be done is to check that the sucker originates above the graft union of the tree in question. That is not always easy to see on an older tree. The sucker must also be stout (around pencil thickness) and healthy.

Height of graft

The higher the graft is introduced, the greater the influence of the rootstock on the new plant. This means that grafting the scion higher up a dwarfing rootstock will produce a more pronounced dwarfing effect than if grafting is done lower down.

Suckers from fruit trees can be suitable as scions

The most important rootstocks for fruit trees and bushes

Rootstock	Vigour	Comments
For apple (*Malus domestica*):		
M 11	very vigorous	
MM 109	very vigorous	
MM 111	vigorous	
M 4	vigorous	v.g. compatibility
M 7	medium vigorous	
MM 106	semi-dwarfing	
M 26	semi-dwarfing	v.g. diameter growth
J 9	dwarfing	staking
Pajam	dwarfing	staking
M 9	dwarfing	staking
M 27	very dwarfing	for tubs
For pear (*Pyrus communis*):		
pear seedling	very vigorous	
OHF 333	vigorous	resistant to fire-blight
Quince A	dwarfing	double working often nec.
Quince C	dwarfing	double working often nec.
For quince (*Cydonia oblonga*):		
Sorbus aucuparia	vigorous	
Quince A	dwarfing	usual rootstock
Quince C	dwarfing	
For medlar (*Mespilus germanica*):		
Sorbus aria	vigorous	
Crataegus monogyna	semi-dwarfing	
Quince A (also for loquat)	medium	also tubs
Quince C (also for loquat)	medium	also tubs
For plum, mirabelle... (*Prunus domestica*):		
Myrobalan group	very vigorous	
INRA GF 8/1	very vigorous	
Brompton	vigorous	
INRA 2	vigorous	
Myruni	medium vigorous	
St Julien d'Orleans	semi-dwarfing	
INRA 655/2	semi-dwarfing	

For peach, nectarine *(Prunus persica)*:

Peach seedling	vigorous	
Almond/peach hybrids	vigorous	
Brompton	vigorous	
Sloe	semi-dwarfing	good for tubs
St. Julien d'Orleans	semi-dwarfing	
Pixi	dwarfing	not with all varieties

For apricot, almond *(Prunus armeniaca, Prunus dulcis)*:

Almond seedling	very vigorous	for dry soils
Myrobalan group	very vigorous	
Apricot seedlings	vigorous	susceptible to disease
Brompton	vigorous	
INRA 2	vigorous	
Myruni	medium vigorous	
INRA 655/2	semi-dwarfing	
St Julien d'Orleans	semi-dwarfing	
Pixi	dwarfing	not with all varieties

For sweet cherry *(Prunus avium)*:

Prunus avium	very vigorous
F 12/1	vigorous
Colt	semi-dwarfing
Maxma Delbard 14	semi-dwarfing
Weiroot	dwarfing
Gisela	dwarfing

For sour cherry *(Prunus cerasus)*:

Prunus avium	vigorous
F 12/1	vigorous
Prunus mahaleb	semi-dwarfing
Maxma Delbard 14	semi-dwarfing
Colt	semi-dwarfing
Gisela	dwarfing

For walnut *(Juglans regia)*:

Juglans regia	very vigorous
Juglans nigra	medium vigorous

For blackcurrant, gooseberry *(Ribes* sp.*)*:

Ribes aureum	for tree-form plants
Ribes divericatum	for tree-form plants

Selected rootstocks for ornamental trees and shrubs

Scion	Rootstock
Abies, fir	*Abies alba, Abies nordmanniana*
Aesculus, horse chestnut	*Aesculus hippocastanum*
Betula, birch	*Betula verrucosa, Betula papyrifera*
Cedrus, cedar	*Cedrus deodara*
Chaenomeles, Japanese quince	*Cydonia oblonga*
Chamaecyparis, false cypress	*Chamaecyparis lawsoniana*
Citrus, species and varieties	*Poncirus trifoliata, Citrus aurantium*
Clematis hybrids	*Clematis vitalba*
Cotoneaster, cotoneaster	*Sorbus aucuparia, Crataegus monogyna*
Fagus, beech	*Fagus sylvatica*
Ginkgo biloba	*Ginkgo biloba*
Hamamelis, witch hazel	*Hamamelis virginiana*
Juniperus, juniper	*Juniperus communis, Juniperus virginiana*
Laburnum, laburnum	*Laburnum anagyroides*
Larix, larch	*Larix decidua, Larix kaempferi*
Liriodendron, tulip tree	*Liriodendron tulipifera*
Magnolia, magnolia	*Magnolia kobus*
Magnolia grandiflora, bull bay	*Magnolia grandiflora*
Olea, olive	*Olea europaea*
Parthenocissus	*Parthenocissus quinquefolia*
Picea, spruce	*Picea abies*
Pinus, leaves in twos (stone pine, Scots pine)	*Pinus sylvestris*
Pinus, leaves in threes and fives	*Pinus strobus*
Rhododendron, evergreen	*Rhododendron ponticum*
Rhododendron, deciduous	*Rhododendron luteum*
Robinia, robinia	*Robinia pseudoacacia*
Salix caprea, willow	*Salix daphnoides*
Syringa, lilac	*Syringa vulgaris, Ligustrum* sp.
Taxus, yew	*Taxus baccata*
Thuja, thuja, arbor-vitae	*Thuja occidentalis*
Tilia, lime	*Tilia platyphyllos*
Tsuga, hemlock	*Tsuga canadensis*
Ulmus, elm	*Ulmus carpinifolia*
Viburnum, viburnum	*Viburnum lantana*

Rootstocks for subtropical fruit trees and bushes

Botanical/common name	Suitable for (top variety)
Acca sellowiana, feijoa	cultivars of this species
Actinidia chinensis, kiwi	cultivars of this species
Annona cherimola, cherimoya	species of this genus, cultivars
Citrus aurantium, Seville orange	Citrus species and cultivars
Diospyros lotus, date-plum	kaki cultivars
Ficus carica, true fig	cultivars of this species
Passiflora sp., passion-flower	*Passiflora edulis*
Persea americana, avocado	cultivars of this species
Pistacia vera, true pistachio	cultivars of this species
Vitis vinifera, grape vine	cultivars of this species
Ziziphus jujuba, jujube	large-fruited varieties

Grafting and budding techniques

There are certain tools and materials which are needed for grafting, and if this task is to be carried out at all frequently, it is worth obtaining these. The most useful tool is a sharp **garden knife** (**grafting knife** or **budding knife**), and a stone for sharpening it. **Secateurs** are also very useful. **Tie material** such as raffia, grafting tape, polythene tape, rubberized strip or rubber budding patches will be needed, and also a coating to paint on to the grafting site, such as grafting wax or pruning compound.

Bud-grafting ("Shield" or "T" budding)
The best time to carry out budding is when the sap is flowing, roughly from early summer to mid-autumn. If it is carried out early, growth will take place in the same year; if it is done from late summer onwards, the bud will still take, but will not make growth until the following year (dormant budding). The second method is particularly common for fruit trees, because it limits frost damage during winter. Budding, as its name suggests, aims to raise a new tree from a single bud of the scion. It is carried out as follows.

A T-shaped cut is made in the stock 10–20cm (4–8in) above the ground, using the budding knife. The blade projection of the knife is then used to

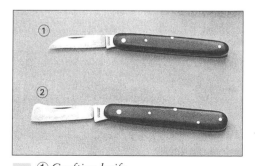

① *Grafting knife*
② *Budding knife*

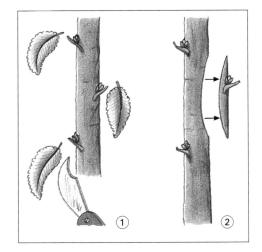

prise open both sides of the incision. This creates a small pocket. A bud is then cut from the scion, working upwards from below, running parallel to the line of the branch. This bud is inserted into the pocket on the stock. The overlap above the bud is trimmed so that it lines up with the horizontal bar of the T-shaped cut in the stock. The bud is tied in place, leaving the bud itself exposed. The site is carefully painted with grafting wax.

It is much easier to use a budding patch rather than tape. The budding patch is simply applied to the site gently, like a plaster, and fastened in place with the attached metal clips. This not only takes much less time but also has the big advantage that it is not necessary to coat with grafting wax.

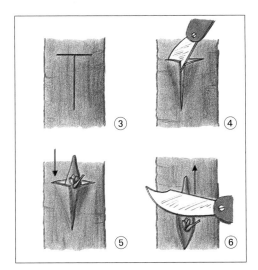

Vigorous rootstocks can stifle the bud, or force it out of the pocket in the rind, however. Another possible problem is that if there is strong formation of callus by the stock at the wound site, this can grow over the bud. Successful union can hardly be expected to result in these cases, so it is advisable not to bud on vigorous stocks too early.

Budding:
① *Remove leaves from scion*
② *Cut out bud for grafting*
③ *T-shaped cut in rind of stock*
④ *Lift rind each side*
⑤ *Insert scion bud*
⑥ *Trim overlap at top*
⑦ *Bud in place*
⑧ + ⑨ *Bud-graft bound:*
⑧ *with raffia*
⑨ *with budding patch*

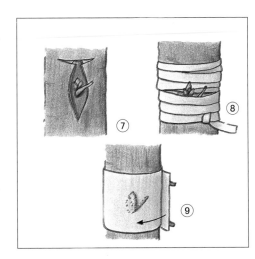

When there are problems of this sort, it also helps to use polythene tape, rather than a budding patch. If polythene tape is used, and fastened quite tightly, this considerably reduces the risk of the bud being forced out or overwhelmed. When budding persimmon *(Diospyros* sp.) for example,

poythene tape should always be used to be on the safe side, although it is more time-consuming. It is also the safest method to choose for early budding on apricot stocks (*Prunus armeniaca*).

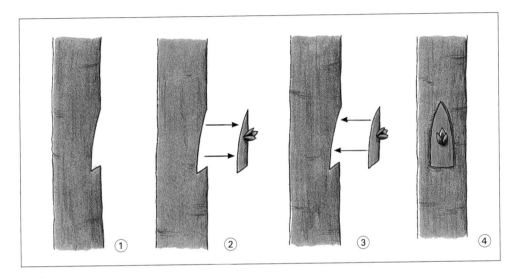

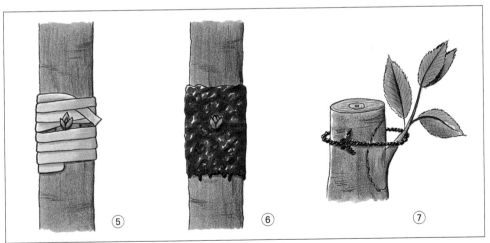

Chip budding:
① Chip cut from stock
② Cutting bud chip from scion
③ Bud chip inserted into stock
④ Bud chip in place on stock

⑤ Binding with rubberized grafting tape
⑥ Sealing with grafting wax: the bud must not be covered!
⑦ Stock cut back, leaving snag; new shoot tied to snag for support

Chip budding

This technique is very similar to "shield" budding; the main difference is that it can be carried out all year round, because it is not necessary to lift the rind. A wedge-shaped chip 1.5–2.5cm (0.6–1in) long is cut out of the stock about 10–20cm (4–8in) from the ground. A similarly sized chip with a bud is cut out of the scion, and pressed into the space left on the stock where the chip was cut out. It is important not to touch any of the cut surfaces with your fingers. The bud chip is then tied in place, and grafting wax applied. Alternatively, overlapping polythene tape can be used to bind the bud chip to the stock; in this case, the grafting wax is not necessary. A budding patch may be used instead.

Once the new shoot is 10cm (4in) long, the polythene tape is carefully slit open at the back of the graft; otherwise it could become ingrown and cause damage, even breakage. If rubber grafting tape was used, there is usually no need to slit this open, because the rubber perishes of its own accord with time from the effects of light, and breaks open.

The risks mentioned above for bud-grafting apply equally to chip budding, especially if budding patches are used.

Splice grafting (whip grafting)

It is best to begin with the following method, if you have not done any grafting before and would like to learn the technique. The large overlap makes for a good chance of creating a successful union. This technique is usually carried out when the sap is not flowing, that is, outside the growing season, in the winter. Pome fruits are often propagated by this method.

The rootstock, which should be about a pencil's thickness, is cut off

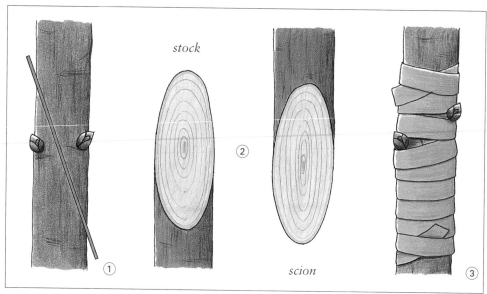

Splice grafting:
① *Prepare by cutting diagonally* ② *View of cut ends of stock and scion*
③ *Stock and scion bound; the union is then sealed with grafting wax*

diagonally about 10–20cm (4–8in) above the neck of the root using a grafting knife. The cut should be 3–6cm (1–2^{1}/$_{2}$in) in length, and opposite a bud. The scion, which should be of the same thickness, is then cut to match. It is trimmed to a length of three to five buds, using secateurs, and placed in position on the cut surface of the stock, lining up the edges to cover it exactly. As with all forms of grafting, you should on no account touch the cut surfaces.

If the scion is slightly thinner than the stock, it should not be placed centrally over it, but at least one side should be lined up to bring the cambium of the one into direct contact with the cambium of the other. Otherwise the two sections will not be able to grow together and unite. The two stems are then simply tied in position and grafting wax applied. If special rubberized grafting tape is used, light-coloured grafting wax should be chosen, as the special rubber decomposes under the effects of the ultra-violet rays in sunlight. It will then be unnecessary to cut open the tape once the first shoot has formed. Dark-coloured grafting compounds do not allow the ultra-violet rays through to the rubber, and so prevent its decomposition. In this case, the tape must be cut open with a sharp knife once the graft has taken, or as the stem begins to thicken.

Whip and tongue grafting

Plants which do not unite easily can be grafted using a particular proven technique which increases still further the area of cambium layer (growth zone) which can overlap. This involves making an additional cut into both the scion and the stock at the point where they are to join. Then, the two sections are not simply placed end to end, but are

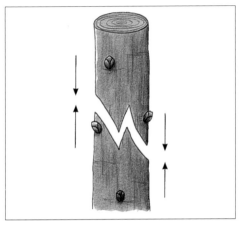

Whip and tongue grafting: stock and scion are pushed together

interlocked. The improved grafting offered by this technique can be used, for example, to raise walnut, sweet chestnut, persimmon and citrus trees.

Inlay grafting

This important technique is not entirely straightforward and requires some practice, but it is useful when the rootstock is significantly thicker than the scion.

In this technique, a wedge-shaped cut is made into the stock. The scion is cut to a matching shape, opposite a bud. The two sections of the graft are then pushed one into the other. After they have been assembled in this way, two half moon-shaped cut surfaces are left protruding above the graft site, on the scion. These stimulate the formation of callus, ensuring that a good quantity of tissue overgrows the site and results in successful union.

A grafting tool developed by an Austrian technical expert has recently appeared, which can prepare the stems for this type of graft precisely, and so ensure a close fit. The tool has performed satisfactorily in tests. This makes it easier for amateurs to carry out inlay grafting.

Side grafting

The main reason for side grafting is to unite a thin scion with a thicker stock. Evergreens and conifers can also be propagated by this method. The stock is prepared during the growing season, when the sap is flowing, by making a T-shaped cut and prising up the rind on both sides, as for budding (see page 56), using the back of the knife or the blade projection to create a pocket. The scion should be about 4–6cm (1½–2½in) long and have around three to four buds. The needles of conifers should be removed

first. Leaves should be cut back to reduce transpiration. A slanting cut about 3cm (1in) long is made at the bottom of the scion stem, which is then pushed into the prepared pocket. The graft is tied and sealed; the cut surfaces at the top of the scion are also sealed. Conifers do not need sealing, because enough resin oozes from the cuts to protect the stem from drying out and from attack by pests.

Veneer grafting

Conifers in particular can also be propagated by this method of grafting. The potted and rooted stock plant is prepared by making a wedge-shaped cut in it, a little above ground level. The lower part of the scion is stripped of its needles, and a slanting cut, as for splice grafting, is made across its end. A second slanting cut is made across the tip of this, to make a wedge which exactly fits into the prepared stock. If the scion is almost

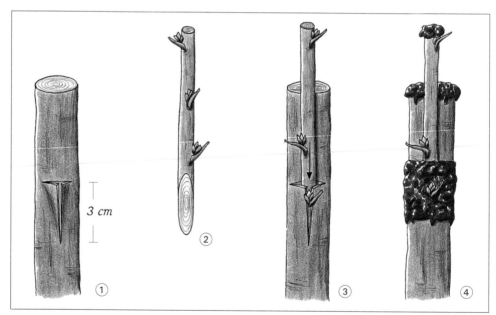

Side grafting:
① T-shaped cut in stock ② Prepared scion ③ Assembled graft, before binding ④ Graft bound and sealed with grafting wax

(but not quite) as thick as the stock, scion and stock should be lined up flush on one side at least, to ensure cambium contact. The two are then bound together, but conifers are not usually sealed with grafting wax when grafted by this method, for the reason given above.

The completed grafts are then laid on the slant in a case lined with damp peat, with their graft site facing upwards, covered with a sheet of glass or plastic film, and placed in a light position at 15–20°C (59–68°F) for four to six weeks, until successful union has taken place. The warm, damp atmosphere in the propagator or greenhouse promotes the union of the graft. The stock is not cut back until the scion shows signs of growth, and then only partially. It is cut back fully in the growing season, in summer.

Side-veneer grafting

This procedure is similar to the veneer grafting technique described above. It is simple to perform, and is chosen when the stock is thicker than the scion, up to around twice the diameter. Although splice grafting could be used where there is only a slight difference in diameter between scion and stock (the cambiums would have to be lined up to coincide, and so make contact on one side at least, in this case), side-veneer grafting is often preferred, because it enables greater cambium contact. The thicker the stock, the more precisely the grafting has to be carried out. A typical use for this technique is in raising pome fruits in winter by grafting.

Top and framework grafting

These two types of grafting involve introducing a scion into an older stock plant. Grafting is often used in southern

Europe to provide new scions for vigorous, strongly growing rootstocks. New varieties can be substituted on older fruit trees to achieve some desired improvement. This type of grafting is carried out for the following reasons:

- The climate does not suit certain varieties of fruit tree. For example, the Cox's Orange Pippin apple may be replaced with one more suited to a harsh, cold climate.
- Storm damage may have broken off part of the crown of the tree. Grafting may be used to restore the shape of the crown as quickly as possible.
- Incompatibility may have resulted in severe damage to the crown at the graft union; the crown may even have broken off. An example of this is a pear grafted on a quince stock, where scion and stock were not sufficiently compatible.
- There may have been physical damage to the union (e.g. gnawing by wild animals) and the grafted variety is to be preserved.
- Where a thick branch has been sawn off, a large wound is left, which needs to be covered with wound tissue quickly. This protects the tree from invasion by harmful organisms. Suitable scions can be grafted into the site, when the branch is removed; the number depends on the thickness of the branch. Once sufficient tissue has overgrown the site, the thinner scions can be cut off.
- The rootstock is so vigorous that callus threatens to overwhelm or force out the newly grafted bud when "shield" budding or chip budding.

Examples are varieties of persimmon on date-plum, or citrus on older rootstocks.

● One fruit variety may fail to bear, because there is no variety which can pollinate it. If, for example, an old-established pear tree in a container on the balcony is not receiving pollen capable of fertilizing it, another variety of pear which is a suitable pollinator can be grafted in.

● The aim may be to produce a vigorous fruit tree, or one which will be in full production quickly. One example would be when there is limited availability of suitable scion material, but a large number of plants true to variety is to be raised as quickly as possible. A tree grafted in this way will provide a large number of buds suitable for grafting in the same or the following year.

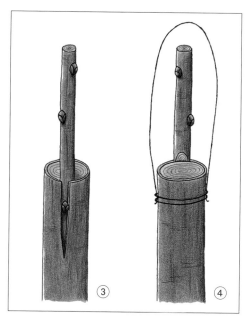

Rind grafting

This method is simple to perform, even without experience. Grafting is done in late spring to early summer, when the sap is flowing in the rootstock. The scions must still be dormant. This means that they need to be cut in advance, and stored in a cool, dark place, for example, a cold store. The scions can equally well

Rind grafting:
① *Stock, cut back, with vertical slit in rind*
② *Prepared scion*
③ *Insert scion into stock*
④ *Rear view of graft, with protective wire*
⑤ *View of graft from above*

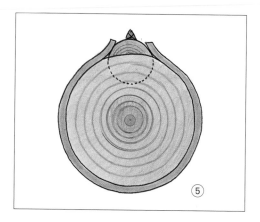

be buried in sand in a shady place in the garden. The spot should be marked, and it should be protected from mice. The branch where the graft is to be inserted is sawn back. The site is then further trimmed with a sharp knife such as a pruning knife. Then a slit about 3 cm (1 in) long is made into the rind of the stock, running lengthwise down it. The back of the knife, or special projection for the purpose, is then used to prise up the rind, lifting it slightly to form a pocket. The scion is prepared by making a slanting cut 3–3.5cm (1–1¹/₂in) long across the bottom, as for splice grafting. The cut should be made opposite a bud. The scion should be 5–8cm (2–3in) long, and have three to five buds. The prepared scion is pushed down into the pocket in the rind until it holds itself in position. A strong raffia tie is put round it, and this must be carefully coated with grafting wax. The cut tip of the scion is sealed likewise. If the cut surface at the base of the scion protrudes a little, leaving a half moon-shaped area exposed above the stump of the branch, this results in a better union, and helps to prevent any possible breakage through wind damage at a later date. Two or more scions are inserted if the stock is thicker; a branch with a diameter of up to 3cm (1in) can normally support one scion, a branch 3–5cm (1–2in) across can support two, and a branch over 5cm (2in) across can support three scions.

A piece of strong wire can be placed over the area of the graft, and held firmly in position, to prevent birds from settling on the freshly inserted scions and damaging the union.

Grafting cacti

There are various reasons for grafting cacti. Some of these succulents have poor root systems, or only grow very unsatisfactorily. Grafting on a vigorous rootstock is a possibility in these cases.

Coloured cactus mutations such as the red cactus Red Head (*Gymnocalycium mihanovicii* var. *fredrichii*) or the yellow cultivar of the peanut cactus (*Chamaecereus silvestrii*), which lack chlorophyll, will only grow when grafted. However, the union can be so low that the plant appears to be on its own roots.

Finally, flowering, fruiting, and so seed production from cacti can be hastened considerably by grafting one type of cactus on another. This can admittedly alter the appearance of the plant, but the species, or the particular cultivar, remains preserved.

All the vigorous species make suitable rootstocks, for example *Cereus peruvianus*, *Eriocereus*, *Opuntia* and *Pereskia*.

Colourful cactus mutations

Flat grafting is carried out from spring into the summer. Suitable gloves need to be worn to protect the hands.

A smooth, horizontal cut is made through the central portion of the stem – that is, neither through the woody lower portion nor through the soft growing region at the tip – to slice off the top of the rootstock. The rings of the vascular bundles typical of most species will now be clearly visible. Any protruding ribs are cut off with a sloping cut, and spines which may interfere with the grafting site are trimmed back. The scion is prepared in the same way, and placed on to the stock with light pressure, working from one side to the other. This prevents air or foreign bodies becoming trapped in the middle, as these could hinder union. The scion is held in place by putting one or two elastic bands over the top of the scion and under the rootstock in its pot. There is a gadget for this purpose, which has a height-adjustable horizontal bar to put constant, even pressure on the top of the scion.

Apical-wedge grafting

There is a choice of various methods of grafting for cacti as there is for woody plants. Apical-wedge grafting is a further method which deserves to be mentioned here. This method makes it possible, for example, to graft the Easter cactus (Rhipsalidopsis) on a vigorous, slender rootstock to create a "cactus tree." The selected rootstock is cut off at the desired height, using a sharp knife, and a slit 2–3 cm ($^3/_4$–1in) deep is made in it, as centrally as possible. A leaf (section) of the scion is prepared by first scraping off the outer surface layer from

Easter cactus (Rhipsalidopsis)

its lower half. The scion is then pressed down into the cleft in the stock. It is held in place with cactus spines or brass pins. The two partners take only two or three weeks to grow together if kept in a light place, in a humid atmosphere at 24–26°C (75–78°F). The spines or pins can then be removed.

Grafting herbaceous plants

Occasionally a report appears in the press that researchers have succeeded in creating a new breed of plant, a cross between a potato and a tomato. Below ground, potatoes grow, and can be harvested come autumn. Above ground, tomatoes ripen on the stem. A fascinating idea! Is it possible? Not quite as described. It is not a newly discovered species, nor is it the result of genetic manipulation, but of two plants of different genera but the same family (the nightshade family, *Solanaceae*) which have been grafted together. The potato (*Solanum tuberosum*) is used as the rootstock, and is raised in a pot in advance. The tomato plant (*Lycopersicon* sp.) is likewise raised in a pot, at the same time. When both are about 40cm (16in) high, they are grafted one on the other. The scion shoot of the one plant is carefully pulled across to the other. Inarching (sometimes known as "grafting by approach") is the method of grafting recommended in this case; with this method, the scion remains attached to the original parent plant until union with the rootstock has taken place. Whip and tongue grafting is also a possibility (see page 59), carried out in such a way that the scion (the tomato) is not detached from its own rootstock until union has taken place (shoot development shows when this has occurred). The potato stem above the union is removed at this stage, along with any other shoots which emerge from the potato. Aftercare of warmth and light should make it possible to harvest potatoes as well as tomatoes from the plant in late autumn. No research has been done to see if the tomatoes and potatoes really do contain

the same nutritional substances as "normal" ones. Production involves a lot of labour for a small yield, so commercial exploitation is not financially viable.

Other herbaceous plants can also be grafted in this way, to promote particular characteristics or to prevent susceptibility to disease. Cucumbers are grafted on the fig-leaved gourd (*Cucurbita ficifolia*) for precisely these reasons, to prevent cucumber wilt (*Fusarium*) and to produce a vigorous plant.

In this case too, grafting is done using the whip and tongue or the inarching method. Scion and stock are each raised in a pot, ready for grafting, with the scion being planted three to five days earlier, to allow for the fact that the gourd grows more quickly.

Grafting by mechanical means and special tools

A quick mention can be made here of mechanical grafting, and the very similar method of grafting using special tools. Apart from the tool for inlay grafting already mentioned (page 59), grafting by mechanical means has established a role for itself in viticulture especially. A special tool is used to make an omega-shaped cut in both stock and scion. The two are then pressed into position on each other, bound and sealed.

The tongued cut for preparing stock and scion is also common. The union produced has great physical strength.

Grafting of herbaceous plants: When the cut regions have united, the grafted tomato (right) is separated from the parent plant. The potato shoot is trimmed back above the union

Protection from pests and diseases

The most effective means of ensuring that seeds and seedlings grow well is to take preventative measures to protect them from anything which might cause disease, and from pests.

Hygiene begins with tools and equipment. These should be clean, well cared for, and as sterile as possible. Self-prepared composts can be a particular problem, harbouring and transmitting fungus and harmful nematodes. Home-made compost should therefore be sterilized by heat (see page 17). If there are any doubts over whether seeds are germ-free or not, they can be dressed (see page 25).

The healthy growth of seedlings and young plants can also be encouraged by providing ideal growing conditions. It is important to make sure that the soil is never wet. Waterlogging can very quickly lead to rotting, especially in the case of semi-ripe cuttings. The plants are then beyond rescue. Ventilation is also necessary. Plants growing under plastic do indeed normally do well, because the humidity is ideal for them, but these same conditions are also ideal for the spores of various harmful fungi to germinate, unless ventilation is carried out regularly.

Damping off disease is a big danger for newly germinated seedlings. A number of soil-borne fungi are often to blame for this. The thin, delicate stem of the plant quickly discolours to grey or black, and the plant subsequently collapses and dies.

Blackleg is essentially the same phenomenon. Some plants are particularly susceptible to it, and others are more resistant. The lychee is one which is particularly susceptible. The large kernels taken from the fruit are planted and kept in warm conditions, and the seedlings generally appear after two to three weeks. When they reach a height of around 10cm (4in), mass disaster often strikes the seedlings. The stems of all the plants turn dark brown or black, and they collapse. The upper section of the plant then either dries up or rots.

Commercial growers treat nursery beds as required with approved fungicides.

Pests can also cause a great deal of damage to young plants.

Slugs are amongst the most harmful. If the plants being raised take their

Damping off, lettuce

fancy, and when slugs are present in sufficient numbers, they can destroy the green parts of every single plant in one night. Attack by slugs and snails does not necessarily mean instant recourse to pellets of poisonous bait, which use metaldehyde as their active ingredient. Putting up barriers to the slugs, or laying out lettuce leaves and the like, under which to trap and collect them, can often be quite adequate. Traps are also available. These attract the slugs by means of particular odorants.

Aphids love tender, young shoots. If an area of young plants is affected by aphids, it is essential to act quickly. Attack by small numbers can probably be dealt with by removing the aphids manually, without recourse to chemical preparations. Washing them off by spraying is not an option, because the plants are too tender, and the ground might become too wet. Plants can also be protected from pests

Vine weevil damage

by covering them with fleece or other special coverings.

Vine weevils are extremely harmful, and can cause a great deal of damage to plants by eating away at them. Their larvae are at least as serious a pest; they are found under the soil, where they eat the fine roots. Nematodes (eelworms) cause similar damage. A helpful tip here is to plant African marigolds (*Tagetes*) amongst the young plants. Voles can cause widespread damage, for instance, wherever seeds are sown.

Birds can be useful as well as harmful. On the one hand, they get rid of a number of pests but, on the other, they enjoy eating various seeds and seedlings.

Plant protection tips

Hygiene is essential in all aspects of plant propagation. The use of sterilized composts will keep fungi and nematodes, the worst enemies of young plants, at bay. If fungal infections do occur, it is important to provide frequent and thorough ventilation, and if necessary to use an approved proprietary fungicide.

Slugs should be trapped and removed at once; slug traps may also be used.

Where there are nematodes in the soil, it is not possible to raise any young plants. A complete change of soil is the only solution here.

Vine weevils can be collected at night when they come out to feed, but their greedy larvae will be busy destroying the roots. Here too, only a complete soil change or an approved pesticide can provide any real solution.

Woody plants

White lilac *(Syringa vulgaris)*

There are various methods of propagating trees and shrubs. Both seed and vegetative propagation are possible.

Deciduous shrubs

Native shrubs can be propagated by seed. The time to sow is normally dictated by what happens in the wild. The seeds of many species ripen in autumn or winter, and fall to the ground, where they are covered over. They germinate the following spring. They will by then have been exposed to the elements for nearly six months, which usually breaks the seed's dormancy. The seedling develops well, and begins to put out shoots, if it finds suitable conditions. But of course it is not possible to carry out propagation which is regulated in this way. Only a fraction of the seeds would in fact emerge. A large number will be food for animals, and yet others will not find the conditions they need to germinate.

The usual method is therefore to collect the seeds of the trees or shrubs in autumn, clean them, and store them appropriately, which may involve stratifying them (see page 26). The prepared seeds can, of course, be sown on the spot; otherwise this is done in early spring.

Most ornamental shrubs are propagated by hardwood cuttings (see page 36), using the middle section of one-year-old mature stems. The best time to take these is early winter, before the plant has been exposed to severe frosts. Once the cuttings have been taken, and the two ends clearly distinguished from each other, they are stored in a cool place with high humidity.

Cuttings are inserted in spring, after the worst of the frosts in a humus-rich, free-draining soil. The use of rooting compounds can encourage root formation in most species.

Ornamental shrubs which can be propagated by hardwood cuttings

Cornel (*Cornus*), deutzia (*Deutzia*), forsythia (*Forsythia*), privet (*Ligustrum*), honeysuckle (*Lonicera*), mock orange (*Philadelphus*), flowering currant (*Ribes*), willow (*Salix*), spiraea (*Spiraea*), snowberry (*Symphoricarpos*), lilac (*Syringa*), tamarisk (*Tamarix*), weigela (*Weigela*).

Woody plants

Botanical/common name	Cont. layering	Layering	Seed	Suckers	Meristem	Hardwood cuttings	Other stem cuttings	Division	Grafting	Root cuttings	Remarks
Abelia, abelia			x				x				needs warmth
Abies, fir			x				x		x		
Acanthopanax, spiny ginseng			x			x	x			x	
Acer, maple			x				x		x		
Actinidia chinensis, kiwi			(x)				x		x		needs warmth
Akebia, chocolate vine, akebia	x		x								
Amelanchier, service-berry			x					x	x		needs warmth
Aralia, aralia			x	x					x	x	needs warmth
Araucaria, Norfolk Island pine			x								needs protection
Arbutus, strawberry tree			x				x				
Aronia, chokeberry			x					x	x		
Arundinaria, bamboo			(x)					x			
Aucuba, Japanese laurel			(x)				x		(x)		
Betula, birch			x						x		
Buddleja, buddleia			(x)			x	x		x		
Buxus, box							x				in propagator late autumn
Calluna, heather, ling	x		x				x				
Camellia, camellia			(x)				x				needs warmth
Camellia sinensis, tea			x				x				needs warmth
Carpinus, hornbeam			x						x		
Castanea, edible chestnut			x						x*		*needs protection
Cedrus, cedar			x					x*	x*		*indoors
Cercis, Judas-tree			x						x		hot water on to seeds
Chamaecyparis, false cypress			x					x	x		
Choisya, Mexican orange blossom			x				x				needs warmth
Clematis, clematis			x						x		
Colletia, colletia			x				x				needs warmth
Cornus, dogwood		x	x			x	x		x		

Woody plants

Botanical/common name	Cont. layering	Layering	Seed	Suckers	Meristem	Hardwood cuttings	Other stem cuttings	Division	Grafting	Root cuttings	Remarks
Corokia, wire-netting bush			x				x				indoors
Cotinus, smoke tree			x			x	x		x		
Crataegus, hawthorn			x						x		
Cydonia, quince						x			x		
Deutzia, deutzia			x			x	x				
Diospyros, persimmon			x						x		indoors
Elaeagnus, oleaster		x	x				x		x		
Erica, heath, bell heather			x				x				
Euonymus, spindle tree			x				x		x		
Fagus, beech			x						x		
Ficus carica, fig			(x)			x	x		(x)		needs warmth
Forsythia, forsythia	x		(x)				x				rooted in water
Fraxinus, ash			x						x		
Gaultheria, gaultheria	(x)		x				x		x		
Genista, broom			x				x		x		
Ginkgo, ginkgo, maidenhair tree			x			(x)	(x)		x		
Gleditsia, honey locust			x						x		
Halesia, silver bell, snowdrop tree	x		x								
Hamamelis, witch hazel	(x)		x						x		stratify 2 yrs
Hedera, ivy							x		(x)		
Hippophae, sea buckthorn			x			x					clean seeds
Hydrangea, hydrangea			x			x	x		x		
Ilex, holly			x				x				
Jasminum, jasmine			x				x				hardy according to species
Juglans, walnut			x						x		
Juniperus, juniper	x		x				x*		x		*under glass
+ *Laburnocytisus*									x		
Laburnum, laburnum			x			x			x		

Woody plants

Botanical/common name	Cont. layering	Layering	Seed	Suckers	Meristem	Hardwood cuttings	Other stem cuttings	Division	Grafting	Root cuttings	Remarks
Larix, larch			x				(x)		x		
Liquidambar, sweet gum		x	x								needs protection
Liriodendron, tulip tree		x	x						x		
Lonicera, honeysuckle	x		x			x	x				
Lycium, box thorn			x			x					
Magnolia grandiflora, magnolia		x	x				x		x		needs warmth
Magnolia sp., magnolia		x	x				x		x		
Malus, apple			x				(x)		x*		*fruit varieties
Mespilus, medlar			(x)						x		
Metasequoia, dawn redwood			x				x				young plants in propagator
Morus, mulberry			x						x*		*indoors
Myrica, bayberry		(x)	x								
Nothofagus, southern beech		x	x								
Osmanthus, fragrant olive			x				x				needs warmth
Paeonia, peony			x					x	x		protect
Parthenocissus, incl. Vir. creeper			x			(x)	x		x		
Paulownia, paulownia			x							x	needs warmth
Pernettya, pernettya			x			x*	x*				*indoors
Philadelphus, mock orange			(x)			x	x				
Phyllostachys, bamboo					x			x			needs protection
Picea, spruce			x				x		x*		*under glass
Pieris, pieris			x				x				needs protection
Pinus, pine			x						x		
Pinus pinea stone pine			x								needs warmth
Platanus, plane			x			x			(x)		
Polygonum, knotweed						x	x				very easy to root
Poncirus, Japanese bitter orange			x				x		(x)		needs protection
Populus, poplar			x			x			x		

Woody plants

Botanical/common name	Cont. layering	Layering	Seed	Suckers	Meristem	Hardwood cuttings	Other stem cuttings	Division	Grafting	Root cuttings	Remarks
Prunus, cherry, plum, peach, nectarine, apricot, mirabelle			x		x	(x)	(x)		x*		*fruit varieties
Pseudotsuga, Douglas fir			x						x		
Pyracantha, firethorn			x				x				
Pyronia, pear-quince cross									x		
Pyrus, pear			x		x				x*		*fruit varieties
Quercus, oak			x				(x)		x		
Rhamnus, buckthorn	x	x					x		x		
Rhododendron, azalea	x	(x)			x		x	x	x		
Rhus, sumach			x	x						x	
Ribes, currant, gooseberry	x	(x)				x	x		x		
Robinia, robinia, false acacia			x						x	x	
Rosa, rose			x			x	x		x		
Rosmarinus, rosemary			x				x				needs warmth
Ruscus, butcher's broom			x					x			
Salix, willow	x	x				x	x		x		easy to root
Sambucus, elder			x			x	x	(x)			
Sciadopitys, umbrella pine			x				x		x		slow-growing
Sequoia, coast redwood			x								needs protection
Sequoiadendron, giant redwood			x						(x)		needs protection
Skimmia, skimmia			x				x				
Sophora, kowhai			x				x		x		
Sorbus, mountain ash, whitebeam, service tree			x						x		
Spiraea, spiraea		(x)				x	x				do not cover seed
Symphoricarpos, snowberry		(x)					x				
Syringa, lilac		(x)					x	x	x		
Tamarix, tamarisk						x	(x)				

Woody plants

Botanical/common name	Cont. layering	Layering	Seed	Suckers	Meristem	Hardwood cuttings	Other stem cuttings	Division	Grafting	Root cuttings	Remarks
Taxodium, swamp cypress			x						x		
Taxus, yew			x				x		x		lime-loving
Thuja, thuja, arbor-vitae			x				x		(x)		moist soil
Tilia, lime	(x)		x						x		
Tsuga, hemlock			x					(x)	(x)		
Ulmus, elm	x		x					x	x		
Vaccinium, blueberry			(x)				x	x			
Viburnum, Guelder rose	x		x				x		x		sometimes needs warmth
Vitis, grape vine			(x)			(x)			x*		*fruit varieties
Weigela, weigela						x	x				
Wisteria, wisteria	x		(x)						x		
Zanthoxylum, Japan pepper			x								needs warmth
Zelkova, zelkova	(x)		x						x		
Ziziphus, jujube			x					x	x		needs warmth

Conifers

Spring sowing is to be preferred for conifers. The ground should be dug and prepared in autumn. Before sowing, the surface of the soil is simply loosened a little, and a little well rotted compost worked in as necessary. The seeds should be sown in rows, allowing plenty of space between them, which makes the later cultivation of the soil much easier. It also makes it easier to estimate the germination capacity of the seeds. That is quite a significant piece of information when raising young plants commercially. The seeds should be sown 2–5cm (³/₄–2in) deep, depending on seed size and external factors. A rule of thumb is to plant at a depth two to three times the thickness of the seed, but at a minimum of 2cm (³/₄in). The reason for this minimum depth is that the surface of the soil dries out very quickly if there are periods of dry weather, and this could harm the seedling.

When sowing the seeds of most conifer species in spring, however, the situation is different. The seeds are

*Yew (*Taxus baccata*)*

*Arbor-vitae (*Thuja occidentalis*)*

only worked into the upper layer of soil, not several centimetres down. The beds are then covered with medium-fine sand. This provides additional protection against drying out and against wind drift. Although many conifers are described as requiring light to germinate, it is not necessary to go to the extent of laying the seeds on the surface of the soil and simply pressing them into place. This turns out to be totally impractical as a way of planting them.

Small quantities of tree seeds can equally well be sown and raised in pots. The young plants are then either planted out or, in the case of more tender ones, kept damp and under glass.

The one-year-old seedlings in autumn – or the cuttings, if propagation is done vegetatively – are grubbed up once mature, and kept in a cold store, or heeled in on a patch of ground, or planted in their intended position.

Spruce (*Picea*) are dug up at the beginning of autumn and transplanted

straight away. This enables them to establish their roots in the same year, and makes them much better able to survive periods of dry weather in spring. The same applies to the arbor-vitae (*Thuja*) and some species of fir (*Abies*). Other conifers, such as larch (*Larix*), the Douglas fir (*Pseudotsuga menziesii*), species of pine (*Pinus*) and yew (*Taxus*) are also planted in spring, as long as they are still dormant. Ginkgo (*Ginkgo biloba*), which can also be counted amongst the conifers (although its leaf shape hardly suggests it), should be planted in spring.

Fruit trees and bushes

Fruit trees are normally propagated either by grafting or by cuttings, hardwood or otherwise. Most fruit trees are raised by grafting. This not only preserves the characteristics of the cultivar but also makes it possible to influence the shape and vigour of the tree.

*Plums (*Prunus domestica*)*

*Redcurrants (*Ribes rubrum*)*

Pome fruits such as apple, pear, quince and medlar can be grafted in winter by splice grafting, or in summer by "shield" budding (dormant budding).

In the case of **stone** fruits, there is a great risk in the winter that the grafted scion may begin shoot development too early, before the union is complete and the scion can be supplied with water and nutrients. It dries up at the base, and makes union impossible. For this reason, stone fruits are "shield" or chip budded in summer. There are exceptions, of course, especially when grafting is done early, and the graft is kept in conditions of high humidity until it is time to line them out. Grafting can be done at other times, using suitable techniques. Winter-grafted trees are lined out when the danger of hard frosts is over, usually in late spring. Until then, the plants are kept in cool conditions. It is also possible to grow them in pots, following grafting in winter.

Certain **fruit bushes** are raised mainly from cuttings (hardwood and other types of stem cutting). The times for taking and inserting cuttings are the same as those given for ornamental shrubs.

Fruit bushes which can be raised from cuttings

Kiwi (*Actinidia arguta, Actinidia chinensis*), chokeberry (*Aronia*), figs (*Ficus carica*), mulberry (*Morus*), Japanese bitter orange (*Poncirus trifoliata*), plum cultivars (*Prunus domestica*), pomegranate (*Punica granatum*), currants (*Ribes* sp.), gooseberries (*Ribes* sp.), Jostaberries (*Ribes* x *nidigrolaria*), elderberry (*Sambucus nigra*), cultivated blueberry (*Vaccinium* sp.), vines (*Vitis*), jujube (*Ziziphus jujuba*), and fruit cultivars such as types of apple, quince, plums and various fruit crosses.

Herbaceous perennials and biennials

Herbaceous perennials, other flowering plants and plants for the balcony are propagated both by means of seed and also vegetatively. The most important means of vegetative propagation is certainly division, and other important options are propagation by stem cuttings, by suckers and by means of root cuttings.

Raising from seed

It is possible to propagate a great many species yourself by seed. This is particularly common with the wild forms, but many cultivated forms are also usually raised from seed. Very hardy species can be sown in the open in winter. Some seeds need a period of cold or frost, followed by rising temperatures, in order to break their dormancy so that they can germinate. Frost is not essential for this, however. The most reliable pregermination treatment is first to soak them in warm water, and then place them in propagating containers in the open, where the low winter temperatures can have their effect. Stratification (see page 26) is often also a possibility. The seeds sown in this way must always be protected from pests.

Biennials should best be sown in summer or late summer, also after the seeds have been harvested. Plants in this category include in particular the familiar pansies (*Viola* x *wittrockiana*), as well as daisies (*Bellis*), foxglove (*Digitalis*), Iceland poppies (*Papaver*) and various species of thistle (*Carduus* sp.).

The seeds of many other herbaceous plants have no need of cold winter temperatures to ensure germination. The time for sowing is less critical for these species. But sowing should be done in time to allow the plant to flower the same year. That normally means sowing

*Foxglove (*Digitalis purpurea)

*Double daisy (*Bellis perennis)

by around mid-spring. Most of these species need fairly high temperatures to germinate, so it is recommended to sow them in pots, and stand the pots in a greenhouse or indoors on a windowsill for a time. When the plants have reached a height of about 10cm (4in), they can be pricked out. They should not be kept anywhere that is too warm or too dark, because this could cause them to become etiolated. They need to be hardened off before planting out as mid-summer approaches (after the late Spring frosts sometimes called the "Ice Saints"). This means gradually accustoming the plants to the colder conditions they will encounter outside.

Propagating by division

Most herbaceous perennials can be propagated true to type by division.

This method has further advantages; it is possible within a very short time to produce new clumps capable of flowering, and which can once more be divided the following year. The division of some species can be done simply by pulling the plant apart with one's hands. An example of this is lily-of-the-valley (*Convallaria majalis*). Others, such as leopard's bane (*Doronicum*), need a

knife, or a pair of strong hands. For older specimens of phlox (*Phlox*), for example, a sharp spade will be needed. The important point is always to cut cleanly. Broken pieces of plant or root should not be left hanging after division; they could rot.

Spring-flowering plants are divided after flowering, and planted straight away in humus-rich soil. They will then grow into fine plants by autumn. Autumn-flowering plants are divided correspondingly later, or (like many perennial grasses and bamboo) before growth begins in spring.

Rejuvenating perennials
Herbaceous perennials are not only divided in order to propagate them but also to refresh older, established plants. The old plant is dug up and divided into a suitable number of smaller sections, which are replanted. This rejuvenation exercise does them good.

Lily-of-the-valley (Convallaria majalis)

Phlox (Phlox drummondii)

Cuttings

Some herbaceous perennials can also be raised from cuttings, though this method is less usual. The cuttings are taken as tip cuttings from growing tips of shoots until mid-summer. Depending on the species and the distance between buds, they should be up to 10cm (4in) long. (See page 32 for further instructions on care of cuttings.)

Achillea (*Achillea*), gypsophila (*Gypsophila paniculata*), penstemon

Root cuttings

Species such as the alpine thistle (*Carlina acaulis*) and perennial poppies (*Papaver* sp.) can be propagated by root cuttings (see page 48).

The cut surfaces of geranium cuttings (*Pelargonium* sp.) should be allowed to dry for a few hours after the cut was made. This is done to prevent possible rotting of the cutting once it has been inserted into the soil.

The subjects should not be set in the open or planted in containers on the balcony until the arrival of mid-summer,

*Sweet violet (*Viola odorata*)*

*Alpine thistle (*Carlina acaulis*)*

(*Penstemon*), and species of violet (*Viola* sp.) are examples of plants which can be propagated in this way.

when all danger of frost is past. The same applies to other tender plants, such as fuchsias (*Fuchsia*).

Herbaceous perennials, summer-flowering plants and herbs

Botanical/common name	Seed	Rhizomes/off-sets	Mainly annual	Greenhouse	Meristem	Suckers	Spores	Cuttings	Division	Rosettes	Sow under glass	Root cuttings	Remarks
Ageratum houstonianum, flossflower	x		x 2–3										
Alcea, hollyhock	4–6												
Anemone x hybrida, anemone									x			x	
Anethum graveolens, dill	x												
Angelica archangelica, angelica	x												
Antirrhinum majus, antirrhinum	1–4		x										
Aquilegia hybrids, columbine	Au												
Aruncus dioicus, goat's beard	Au								x				
Aster sp., Michaelmas daisy	x							x	x				
Astilbe, astilbe									x				
Calendula officinalis, marigold	4–5		x								x		
Callistephus chinensis, China aster	5		x								x		
Caltha palustris, marsh marigold	Au								x				
Campanula sp., bellflower	x								x				
Carex morrowii 'Variegata,' Japanese sedge									Sp				
Chrysanthemum frutescens, marguerite								x					
Chrysanthemum parthenium, feverfew	2–3												
Cleome spinosa, spider flower	x		x 3–4										
Coreopsis grandiflora, tickseed, coreopsis								x	x				
Cortaderia selloana, pampas grass	x								x				protect in winter
Delphinium hybrids, delphinium	x							x	x				
Dianthus barbatus, sweet William	5–7		x										
Dicentra spectabilis, bleeding heart								x					
Digitalis purpurea, purple foxglove	Au												

Spr = spring; Au = autumn; Wi = winter

79

Herbaceous perennials, summer-flowering plants and herbs

Botanical/common name	Propagation												Remarks
	Seed	Rhizomes/off-sets	Mainly annual	Greenhouse	Meristem	Suckers	Spores	Cuttings	Division	Rosettes	Sow under glass	Root cuttings	
Doronicum orientale, leopard's bane	(x)								x				
Dryopteris filix-mas, male fern							x		x				
Echinacea purpurea, coneflower	(x)								x				difficult
Eryngium alpinum, sea holly	x											x	
Euphorbia sp., species of spurge	x							x	x				
Filipendula sp., meadowsweet									x				
Galium odoratum, woodruff	x								x				plant in shade
Gentiana sp., species of gentian	x								x				
Geranium sp., species of cranesbill	x								x				
Gypsophila, baby's breath	x							x					avoid wet
Helianthemum hybrids, rock rose								x					
Helianthus annuus, sunflower	4–5		x										
Helichrysum bracteatum, everlasting flower	4–5		x 3–4										
Helleborus hybrids, Christmas rose	x								x				
Hemerocallis hybrids, day lily									x				
Hosta, hosta	(x)								x				
Hypericum perforatum, St. John's wort	x	x											
Impatiens hybrids, busy Lizzie			x 2–3										20–24°C (68–75°F)
Lathyrus odoratus, sweet pea	4		x 2–4										
Liatris spicata, gay feathers									x				
Ligularia, ligularia	x								x				
Linum narbonense, flax	x							8					

Spr = spring; Au = autumn; Wi = winter.

Herbaceous perennials, summer-flowering plants and herbs

Botanical/common name	Seed	Rhizomes/off-sets	Mainly annual	Greenhouse	Meristem	Suckers	Spores	Cuttings	Division	Rosettes	Sow under glass	Root cuttings	Remarks
Lobelia erinus, lobelia	x		2–4										
Lupinus polyphyllus hybrids, lupin	x								x				
Lychnis chalcedonia, Maltese cross	x								x				
Matteuccia struthiopteris, ostrich fern		x											
Mentha x piperita, peppermint	x	x							x				
Miscanthus sinensis, zebra grass	x			x					x				sunny position
Monarda hybrids, monarda								x	x				
Myosotis sylvatica, forget-me-not	6–7												
Nepeta racemosa, catmint								5	x				avoid wet
Nicotiniana sylvestris, *N. tabacum*, flowering tobacco	x			3									
Osmunda regalis, royal fern							6						
Paeonia lactiflora, peony									Au				
Panax, ginseng	9–3												protect in winter
Papaver orientale, oriental poppy	x											x	
P. somniferum, opium poppy	x												
Pennisetum alopecuroides, Chinese fountain grass									x				
Penstemon hybrids, penstemon	x			2				10					tender
Petroselinum hortense, parsley	x												
Phlox paniculata, phlox								x	x			x	
Pimpinella anisum, sweet Alice	x												
Polygonum sp., knotweed								x	x				
Primula sp., primrose, cowslip	Wi								x		x		
Pulsatilla vulgaris, pasque flower	Au												
Rudbeckia fulgida, coneflower, rudbeckia	x								x				

Spr = spring; Au = autumn; Wi = winter

Herbaceous perennials, summer-flowering plants and herbs

Botanical/common name	Seed	Rhizomes/off-sets	Mainly annual	Greenhouse	Meristem	Suckers	Spores	Cuttings	Division	Rosettes	Sow under glass	Root cuttings	Remarks
Rudbeckia hirta, coneflower	x			3–4									
Salvia, sage	x			3–4				Au					
Saxifraga, saxifrage								x	x				
Sedum, stonecrop, sedum								x	x				roots easily
Sempervivum, houseleek										x			warmth-loving
Solidago hybrids, golden rod									x				
Symphytum, comfrey		x							x				
Tagetes, African marigold	5		x	2–4									
Thunbergia alata, black-eyed Susan	x			3									needs warmth
Thymus vulgaris, thyme	x								x				
Trollius hybrids, globeflower									3–4				
Valeriana officinalis, valerian	x												
Verbascum bombyciferum, mullein	Au												
Veronica, veronica	x								x				sunny position
Vinca minor, lesser periwinkle									x				easy
Viola x wittrockiana, pansy	6–7												
Waldsteinia geoides, waldsteinia									x				
Waldsteinia ternata, waldsteinia				x							x		shade
Yucca filamentosa, yucca	x					x							

Spr = spring; Au = autumn; Wi = winter.

Key
The numbers refer to months of the year:

    2 = early spring
    3 = mid-spring
    5 = early summer
    6 = mid–summer
    10 = late autumn
    2–3 = early–mid-spring

    4–6 = late spring–mid-summer
    1–4 = late winter–late spring
    4–5 = late spring–early summer
    3–4 = mid–late spring
    5–7 = early–late summer
    2–4 = early–late spring
    6–7 = mid–late summer
    9–3 = mid-autumn–mid-spring

Vegetables and herbs

The growing of vegetables in private gardens is widespread. The plants are mainly grown annually from seed. Often, the seeds are sown early, in late winter or spring in a frost-free frame, in the greenhouse or even on a light windowsill. The seedbed needs to be prepared before sowing takes place in spring. It should not be sited where the same species have been grown for many years. Rotating crops prevents the ground from becoming tired, which reduces yields and produces sickly plants.

The ground needs to be dug over during the preceding autumn to loosen and aerate the soil. Compost or well rotted stable manure can be worked in, especially with light, sandy soils. In spring, just the top layer of soil is loosened. The seeds are sown in prepared beds in the desired manner. The ideal width for the beds is about 1m (3ft). If the seedlings are not to be transplanted after they have emerged, the eventual spacing should be considered at the sowing stage. The germination capacity of the seed needs to be taken into account here, or space will be wasted with excessively large gaps between plants. With very fine seed, it is not really possible to keep to the intended spacing. In this case, the fine seeds can be mixed with sand. When the mixture is sown, the seeds are not too overcrowded.

Some species of seed, such as onions and carrots, are available as pelleted seed (see page 23). Although this is more expensive, it can be placed more precisely. The same applies to seed sold

Pelleted seed is available for carrots and onions

in the form of seed bands. Where possible, sowing should be done in rows or drills. This not only makes the later cultivation of the soil easier (see also page 27), but also makes it easier to achieve the desired depth for the seeds, and to cover them more accurately. The rows are set out parallel to the edge of the bed, and a taut length of string used to mark them out.

The row spacing is governed by the type of vegetable being grown. Usually the seed is sown 2–4cm ($^3/_4$–$1^3/_4$in) deep, depending on seed size and soil type; in light soils, it is better to sow a little deeper. After sowing, the seed is covered with soil. This is then firmed down, usually with the back of the rake. The seed is best able to germinate if it is in good contact with the surrounding soil, and if the soil is damp. Once the seedlings have emerged, they are thinned out as necessary for the particular type.

If the seed is sown broadcast, as is usual with very fine seed – the seedlings need to be transplanted several times – the seeds still need to be worked into the ground and lightly firmed down in a similar way.

Some vegetables are sown in groups. Each group consists of around three to five seeds. This practice is common for bush and climbing forms of French beans, and for cucumbers, for example. The seeds may also be soaked first, to increase their capacity to germinate. If this is done, the ground should not be allowed to dry out after sowing, or the advantage will be lost. Species of vegetable and herbs which are particularly tender or require warmth should be sown in pots, trays, or frames

The seeds of French beans are sown in groups

from late winter onwards. After hardening off (see page 29), they are transplanted into the open once the weather is suitable. This can be done, for instance, with tomatoes, cucumbers, courgettes and curly kale. Alternatively, they can be grown on in pots or in the greenhouse. This is done, for example, with melons, the more tender F1 hybrid cucumbers, certain tomato cultivars, aubergines and sweet peppers. Annual herbs, and frost-tender herbs growing in the open, will need to be sown again the following year. Valuable but frost-tender plants such as bay, rosemary and rue should either be grown under glass or in pots, so that they can easily be overwintered.

Sowing vegetables

Common/botanical name	Sowing month	Plant in advance (warm conds)	Poss. to grow as perennial	Grow under glass/plastic	Remarks
Artichoke, *Cynara cardunculus*	2,3	x			
Aubergine, *Solanum melongena*	2,3 (4)	x			
Beetroot, *Beta vulgaris* var. *conditiva*	5,6				sow in open, thin out
Blanching celery, *Apium graveolens*, var. *dulce*	2–4		x		
Broad beans, *Vicia faba*	3,4				
Broccoli, *Brassica oleracea* var. *italica*	3,4/5,6	1			
Brussels sprouts, *Brassica oleracea* var. *gemmifera*	3/4,5	1			
Cabbage lettuce, *Lactuca sativa* var. *capitata*	12–2/4–5	1			
Cape gooseberry, *Physalis peruviana*	2,3	x	x		
Carrot, *Daucus carota*	(3) 4–7				
Cauliflower, *Brassica oleracea* var. *botrytis*	2,3/4,5				
Celeriac, *Apium graveolens* var. *rapaceum*	2,3/5,6	1			
Chicory, *Cichorium intybus* var. *foliosum*	end 5				10: lift, force
Chinese cabbage, *Brassica rapa* var. *pekinensis*	spring				7 onwards: plant out
Courgette, *Cucurbita pepo* var. *giromontiina*	2–4/end 5	1			
Cucumber, *Cucumis sativus*	3–5				end 5 onwards: plant out
Cucumber F1 (greenhouse type), *Cucumis sativus*	2–4			x	
Curled celery, *Apium graveolens* var. *secalinum*	2–4/5,6	1			

Sowing vegetables

Common/botanical name	Sowing month	Plant in advance (warm conds)	Poss. to grow as perennial	Grow under glass/plastic	Remarks
Curly kale, *Brassica oleracea* var. *sabellica*	3,4/5	1			transplant until 7
Endive, *Cichorium endivia*	6				transplant
European Welsh onions, *Allium fistulosum*	3–7				thin out
Florence fennel, *Foeniculum vulgare* var. *dulce*	4,5/6,7	1			
French bean (climbing), *Phaseolus vulgaris* var. *vulgaris*	end 5				sow in groups
French bean (bush form), *Phaseolus vulgaris* var. *nanus*	mid 5	1			sow in groups
Iceberg lettuce, *Lactuca sativa* var. *capitata*	1–4	x			5 onwards: plant out
Kohlrabi, *Brassica oleracea* var. *gongylodes*	12–2/2–5	1			
Lamb's lettuce, *Valerianella locusta*	7–9				lime-loving
Leek, *Allium porrum*	2,3/4,5	1			
Loose-leaf lettuce, *Lactuca sativa* var. *crispa*	(3) 4,5				
Onion, *Allium cepa*	3–5/mid 8				harvest in following year 2
Pak choi, *Brassica rapa* var. *chinensis*	(6) 7				thin out
Pea, *Pisum sativum*	(3) 4,5				
Pumpkin, *Cucurbita pepo*	3,4/end 5	1			2: sow in groups
Radicchio, *Cichorium intybus* var. *foliosum*	5–7				thin out
Radish, *Raphanus sativus* var. *sativus*	10–3/4–9			1	2: sow in open

Sowing vegetables

Common/botanical name	Sowing month	Plant in advance (warm conds)	Poss. to grow as perennial	Grow under glass/plastic	Remarks
Red cabbage, *Brassica oleracea* var. *capitata ruba*	2,3/4,5	1			
Runner beans, *Phaseolus coccineus*	end 5–6				sow in groups
Savoy cabbage, *Brassica oleracea* var. *sabauda*	2,3/4,5	1			
Scorzonera, *Scorzonera hispanica*	3–5				sow in open
Spanish radish, *Raphanus sativus* var. *niger*	(4)5–8				
Spinach, *Spinacia oleracea*	2,3/9			1	2: sow in open
Spring cabbage, *Brassica oleracea* var. *capitata*	2,3/4,5	1			
Sweet melon, *Cucumis melo*	12–2			x	
Sweet pepper, chilli pepper, *Capsicum annuum*	1–5				warmth-loving
Sweetcorn, *Zea mays* convar. *saccharata*	end 5				sow in groups
Swiss chard, spinach beet, *Beta vulgaris* var. *cicla*	4–6				
Tomato, *Lycopersicon lycopersicum*	2–4			x	
Water melon, *Citrullus lanatus*	2–4				warmth-loving
White cabbage, *Brassica oleracea* var. *capitata f. alba*	2,3/4,5	1			

Key

The numbers refer to months of the year:
1 = late winter, 2 = early spring, 3 = mid-spring, 4 = late spring, 5 = early summer, 6 = mid-summer, 7 = late summer, 8 = early autumn, 9 = mid-autumn, 10 = late autumn, 11 = early winter, 12 = mid-winter.

Bulbous plants

Plants which form bulbs, corms, tubers and rhizomes can be propagated both from seed and vegetatively, from parts of the plant. Propagation by vegetative means is both quicker and true to variety, and the simplest methods are by dividing tubers and rhizomes, and by means of bulblets, cormlets, bulbils, and cuttings. Strict hygiene is always important when working. If a knife is used to cut a diseased section of a plant, the knife can transfer the disease to healthy tissue. If in doubt, then, the knife should be disinfected. The best way of protecting larger cut surfaces from infection is by dusting them with powdered charcoal.

Plants which are to be propagated during dormancy need to have ripened fully before this is done. The usual sign of this is when the foliage has died right back. The dying foliage of plants which form bulbs or corms may often look unsightly, but it should not be cut off after flowering. It must be allowed to remain on the plant, because the substances in the leaves which will make up the plant's food stores and metabolites need to be transported back to the root region, which then has to strengthen and mature. This is necessary to ensure satisfactory results when propagating these plants.

Division or separation of tubers

A number of plants can be propagated easily by cutting or separating their tubers, once the plant has died down. The tubers of the winter aconite (*Eranthis*) or the familiar potato (*Solanum tuberosum*), for example, are cut into pieces, each of which must have at least one eye (bud). A young potato plant can often even be raised from nothing more than a suitable piece of skin peeling from the potato. Dahlias (*Dahlia* hybrids) form several tubers during the growing season, and these are carefully separated. The bud must not be allowed to break off. Potatoes can also be propagated in this way.

Division of rhizomes

Plants such as the iris (*Iris*), lily-of-the-valley (*Convallaria*), red-hot poker (*Kniphofia*) and asparagus (*Asparagus*) are propagated by division of their rhizomes. The plant is lifted in late summer, and the rhizome cut into pieces.

Each section of rhizome must have at least one growth bud if it is to produce a new shoot, and so a new plant.

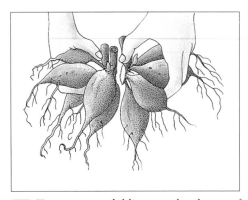

To propagate dahlias, cut the cluster of tubers into portions. Each must have at least one growth bud

The bulblets can be pulled away from the parent plant easily, and planted

Bulblets

Plants which form bulbs also very frequently develop bulblets – small (or even quite large) daughter bulbs, which form in greater or lesser number around the base of the original bulb. The daffodil (*Narcissus*), the tulip (*Tulipa*), and the common onion (*Allium cepa*) develop a few, large bulblets, whereas scilla (*Scilla*), fritillaries (*Fritillaria meleagris*), the blood lily (*Haemanthus*), camassia (*Camassia*) and hyacinths (*Hyacinthus*) form several small bulblets. In the case of hyacinths and the crown imperial (*Fritillaria imperialis*), it is even possible to encourage these small bulbs to form. To do this, the base of the bulb is scored across with a knife several times before planting. The bulb can even be planted upside down, with its basal plate uppermost. This further promotes the formation of bulblets.

Bulblets are separated from the parent plant once the foliage has died down, and they are then planted immediately. However, it is possible to store them instead in dry conditions, and plant them later. Bulblets may not always flower the following year; it may take two or three years for them to reach that stage.

Cormlets

Cormlets resemble bulblets, the difference being in the structure of their tissue. Bulbs consist of several layers of fleshy scale leaves, in which water and nutrients are stored. These surround the stem and the embyronic leaves and flowers, and are held together by the basal plate of the bulb.

In corms, the scale leaves have in effect grown together to form a tissue that serves the same purpose. It contains the shoot. Just as bulbs form bulblets, so

Cormlets develop on crocus corms

89

plants such as the crocus (*Crocus*), buttercup (*Ranunculus*), erythronium (*Erythronium*), autumn crocus (*Colchicum*) and sparaxis (*Sparaxis hybrids*) form either cormlets (also called cormels) or clusters of tubers.

If plants are being propagated by means of these cormlets, generous feeding of the parent plant during the growing season is necessary, and the corm must be allowed to ripen completely. Otherwise, the cormlets will be very small, and vulnerable to any shortcomings in their growing conditions.

Bulbils

Propagation from bulbils is rather different. Some lilies form tiny bulbs called "bulbils" in the axils of their leaves, or bulblets on sections of underground stem. The bulbils or bulblets are carefully removed in autumn, and stored in peat or propagating compost until spring. Then they are either grown on in pots, or planted in situ straight away.

Cuttings

Some cultivars can be raised true to type and quite easily by cuttings. Commercial growers raise dahlias in this way on a large scale. The tubers are forced during the winter months.

Once the stems are 10–20cm (4–8in) high, the cuttings can be taken (taken off at the base, with a heel) and inserted, and grown on under glass. The stock plant is not deprived of every single shoot, so that it can still recover.

Lily bulbils

Seed

Many species can also be propagated from seed. This may sometimes involve losing the particular characteristics of a cultivar. The seed is harvested in summer or autumn, cleaned, dried and then stored in bags or lidded glass jars. The seeds are sown in spring or summer, either in a frame, or directly into the open. Some species actually need frost or chilling to germinate. These should be sown in autumn, or stratified (see page 26). The plants often show little development in the first year, and it often takes two years or more before they flower again.

Bulbous plants

Botanical/common name	Seed	Leaf cutting	Cormlet/tuber	Bulblet	Offset	Division/tubers	Offset/tubers	Division/rhizomes	Cuttings	Division	Root tuber	Remarks
Acidanthera, gladiolus			x									late spr/early sum
Acorus, sweet flag										x		
Agapanthus, agapanthus	(x)									x*		*early/mid-spring
Allium, onion, allium	x			x*								*autumn
Alstromeria, Peruvian lily	x**						x*					*spring, store cool
Anemone, anemone	x									x	x	autumn, spring
Arum, arum	x	x										
Begonia, begonia	x					x			x			
Belamcanda, belamcanda	x									x		
Bletilla striata (orchid)										x		
Camassia, camassia	x			x								
Canna, canna	(x)						x*					*winter, spring
Chionodoxa, glory-of-the-snow	x			x								
Clivia, clivia					x							spring, summer
Colchicum, autumn crocus	x	x										
Convallaria majalis, lily-of-the-valley									x			autumn–spring
Crinum, crinum				x								
Crocus, crocus	(x)		x*									*summer
Crocus sativus, saffron			x									
Cyclamen, cyclamen	x											
Dahlia, dahlia	x					x			x			
Dracunculus, dragon arum	x						x					
Eranthis, winter aconite	x					x						
Eremurus, foxtail lily	x									x*		*summer
Erythronium, erythronium			x									summer, plant imm.
Eucomis, pineapple flower	x			x*								autumn
Freesia, freesia	(x)		x									
Fritillaria imperialis, crown imperial				x								late summer

spr = spring, sum = summer, imm = immediately

Bulbous plants

Botanical/common name	Seed	Leaf cutting	Cormlet/tuber	Bulblet	Offset	Division/tubers	Offset/tubers	Division/rhizomes	Cuttings	Division	Root tuber	Remarks
Fritillaria meleagris, snake's head fritillary	(x)			x*								*autumn
Galanthus, snowdrop	(x)			x								
Gladiolus, gladiolus			x							x		
Haemanthus, blood lily				x								
Helianthus tuberosus, Jerusalem artichoke						x						autumn–spring
Hyacinthus, hyacinth				x								
Ipheion, ipheion				x								
Iris, iris	x**			x*								*sum–aut, **sum
Kniphofia, red-hot poker, torch lily								x				
Leucojum vernum, snowflake	x			x								
Liatris, gay feathers	x									x*		*spring
Lilium, lily	x			x								also bulbils
Muscari, grape hyacinth	x			x								
Narcissus, daffodil, narcissus				x								
Oxalis, wood sorrel	x			x								
Puschkinia, puschkinia	x**			x*								*early aut, **sum
Ranunculus asiaticus, Persian buttercup	(x)*			x								late autumn–mid-spr under glass
Scilla, scilla	x			x								
Sinningia, gloxinia		x										summer
Sparaxis, sparaxis	x		x									
Sternbergia, sternbergia				x								mid-summer
Tigrida, tiger flower	x*			x								*spring
Tritelia, tritelia	x*			x								*autumn
Tulipa, tulip	(x)*			x								*wild tulips

spr = spring, sum = summer, aut = autumn

Patio and indoor plants

A patio plant is generally one being grown in quite a large container; often it is not hardy, and so needs to be overwintered in a suitable place indoors. Using this as a definition, plants raised from the seeds of exotic fruits, which are becoming more and more popular, could also be placed in this category.

An indoor plant is usually one grown for its flowers or foliage in a small pot, situated in a warm place in the home. It is usually of tropical origin.

A good many patio and indoor plants are propagated vegetatively, because in this way the known characteristics of the parent plant are preserved, and propagation can be carried out quickly. Cuttings are the means of propagating Schefflera (*Schefflera*), the chenille plant (*Acalypha*), golden trumpet (*Allamanda*), the wax plant (*Hoya*), angel's wings (*Caladium*), species of fig (*Ficus benjamina, Ficus lyrata, Ficus elastica, Ficus carica* and others), oleander (*Nerium*), pomegranate (*Punica*), Laurustinus (*Viburnum tinus*), angel's trumpet (*Datura*), and the lagerstroemia or crape myrtle (*Lagerstroemia*).

Grafting is the method used to propagate most species of citrus, as well as the kaki or persimmon (*Diospyros kaki*), the loquat (*Eriobotrya japonica*), the large-fruited jujube (*Ziziphus jujuba*), the white mulberry (*Morus alba*), with its delicious, large fruits, and various acacias (*Acacia*).

Seed too is used to propagate various species. Amongst these are many members of the Leguminosae family, such as the coral tree (*Erythrina crista-galli*), the silk tree (*Albizia*), Barbados pride (*Caesalpinia*), the fairy duster (*Calliandra*) and sesbania (*Sesbania*), as well as feijoa (*Acca sellowiana*) – though particular cultivars of this will be

Schefflera can be propagated by means of cuttings

Propagating pineapple
To propagate pineapple (*Ananas comosus*), the tuft of leaves is cut off, cleaned of the fruit flesh, and the lower leaves removed. The resulting rosette of leaves is then planted in well draining compost. Aftercare involves keeping it at 25–30⁰C (77–86⁰F) in conditions of high humidity. The rosette will then soon begin to show signs of growth, and become an attractive plant.

The loquat (Eriobotrya japonica) is propagated by grafting

propagated by cuttings or grafting.

It is also possible to propagate from seed the jacaranda (*Jacaranda*), ornamental bananas (*Ensete ventricosum*, *Musa* sp.), *Radermachera*, palms such as date palms (*Phoenix* sp.), the fan palm (*Washingtonia*) and the Japanese sago palm (*Cycas revoluta*). It is very important when propagating these plants to maintain the warmth of the growing medium at 20–24⁰C (68–75⁰F), and generally high humidity.

The same can be said of growing plants from the seeds of exotic fruits. The large, rounded seed of the avocado is removed from the fruit and fixed in position above a container filled with water. After four to six weeks, the first roots appear, and once this has happened, the seed is potted up. It soon grows into an attractive plant. The seeds of the lychee need to be thoroughly cleaned and, as far as possible, disinfected before planting in propagating compost; otherwise they tend to rot easily. The young seedlings have to be dealt with extremely carefully.

The compost should be as sterile as possible, because soil-borne fungi could quickly destroy the young plants. Some very respectable climbing plants can also be raised from the cleaned seeds of various varieties of passion-fruit, if they are planted in a warmed substrate at 22–24⁰C (71–75⁰F). The whole fruit of the chayote or vegetable pear (*Sechium edule*) is simply planted 2cm (³/₄in) deep in a container filled with soil. The chayote soon shows signs of growth at room temperature. A high proportion of the cherimoya's many seeds will germinate, if they are sown fresh. The plants can be grown indoors. Once they get bigger, they need to be placed in a conservatory, and outside in summer. The seeds of the guava (*Psidium guajava*) also germinate reasonably. If the resulting plant is kept under glass, except for being planted outside in fertile soil

Avocado seedling

during the best of the summer months, it could even flower after three years, and with a little luck, perhaps even fruit. The fruits borne by these seedlings are not very large, but the scent of the ripe fruit is intense and very pleasant. Similar success can be had with the seeds of star fruit (*Averrhoa carambola*), papaya (*Carica papaya*), tamarillo (*Cyphomandra*) and a number of other exotic fruits.

The plants grown from the pips of citrus fruit can be used as rootstocks.

Patio and indoor plants which can be propagated from seed

The fruits of plants marked with * are sometimes available in the shops, and seeds can be extracted from them.

Abutilon (*Abutilon*), Mimosa, acacia (*Acacia*), copperleaf, Jacob's coat (*Acalypha wilkesiana*), feijoa (*Acca*)*, agapanthus (*Agapanthus*), silk tree (*Albizia*), cherimoya (*Annona cherimola*)*, flamingo flower (*Anthurium scherzerianum*), peanut (*Arachis hypogea*)*, Norfolk Island pine (*Araucaria*), ardisia (*Ardisia crispa*), jackfruit tree (*Artocarpus heterophyllus*)*, Japanese laurel, spotted laurel (*Aucuba*), star fruit, carambola (*Averrhoa*)*, Barbados pride (*Caesalpinia*), bottlebrush (*Callistemon*), camellia, tea (*Camellia*), caper (*Capparis spinosa*), papaya (*Carica papaya*)*, Natal plum (*Carissa*), cassia (*Cassia*), catalpa (*Catalpa*), cestrum, bastard jasmine (*Cestrum*), dwarf fan palm (*Chamaerops*), Mexican orange blossom (*Choisya*), species of citrus (*Citrus*)*,

coconut palm (*Cocos*)*, coffee (*Coffea*), sweet melon (*Cucumis melo*)*, Japanese sago palm (*Cycas*), papyrus (*Cyperus papyrus*), tree tomato, tamarillo (*Cyphomandra*)*, angel's trumpet (*Datura*), kaki, ebony, persimmon (*Diospyros*)*, durian (*Durio zibenthinus*)*, ornamental banana (*Ensete ventricosum*), loquat (*Eriobotrya*)*, coral tree (*Erythrina*), eucalyptus (*Eucalyptus*), pitanga (*Eugenia uniflora*), species of fig (*Ficus*)*, species of fuchsia (*Fuchsia*), jacaranda (*Jacaranda*), jasmine (*Jasminum*), crape myrtle, lagerstroemia (*Lagerstroemia*), lantana (*Lantana*), laurel (*Laurus*), litchi, lychee (*Litchi chinensis*)*, macadamia nut (*Macadamia tetraphylla*)*, magnolia (*Magnolia*), mango (*Mangifera indica*)*, ornamental banana species (*Musa* sp.), sacred bamboo (*Nandina*), rambutan (*Nephelium lappaceum*)*, ochna, Mickey-Mouse plant (*Ochna*), olive (*Olea*), prickly pear, Indian fig (*Opuntia ficus-indica*), passion-flower (*Passiflora*), avocado (*Persea americana*)*, date palm (*Phoenix*)*, mastic tree, pistachio (*Pistacia*), pittosporum (*Pittosporum*), guava (*Psidium guajava*)*, pomegranate (*Punica*)*, chayote (plant the whole fruit) (*Sechium edule*), various solanaceous plants (*Solanum*), African hemp (*Sparmannia*), bird-of-paradise flower (*Strelitzia*), chocolate tree (*Theobroma cacao*), yellow oleander (*Thevetia*), glory bush (*Tibouchina*), windmill palm, Chusan palm (*Trachycarpus*), fan palm (*Washingtonia*), jujube (*Zizyphus*).

Bonsai, hydroculture, ferns, meristem

This chapter will describe some special features of bonsai and of plants destined for hydroculture. It goes on to discuss the propagation of ferns, and to outline propagation by meristem culture. This last is not, however, a method which the amateur would expect to be carrying out him or herself.

Bonsai

Plants which grow in the open, or in containers, are not essentially different from those grown as bonsai. It is normally simply the growing technique which produces the dwarfed appearance, and the tendency in many cases for the plant to assume its mature form sooner than it would naturally. In this chapter, the raising of plants for bonsai is discussed, but the styles, techniques and care lie outside the scope of what can be said here.

A number of native trees are suitable for raising from seed as bonsai. Such plants can normally live to an age of at least fifty years. They are grown in containers. Two years from planting, the main or tap root is cut back. This encourages the growth of fibrous roots. In the following year, the first pruning of shoots and branches to achieve the desired style begins.

The trunk of this bonsai spruce, and the degree of development of the tree's overall shape, suggest that this specimen is fairly old.

Bonsai plants to raise from seed
Field maple (*Acer campestre*) and other species of maple, alder (*Alnus*), weeping birch (*Betula pendula*), European hornbeam (*Carpinus betulus*), beech (*Fagus sylvatica*), Hong Kong kumquat (*Fortunella hindisii*), ash (*Fraxinius excelsior*), common juniper (*Juniperus communis*), European larch (*Larix decidua*), apple (*Malus*), mulberry (*Morus*), olive (*Olea europea*), spruce (*Picea*), pine (*Pinus*), pomegranate (*Punica granatum*), pear (*Pyrus*), common oak (*Quercus robur*), mountain ash (*Sorbus aucuparia*), service tree (*Sorbus domestica*), yew (*Taxus baccata*), small-leaved lime (*Tilia cordata*).

The raising of these plants can be speeded up by lining out the two-year-old plants in the open, and carrying out regular pruning for three years. Only then are the plants dug up, the required trimming carried out, and the subjects replanted in shallow bonsai containers.

Hydroculture

Where plants intended for hydroculture are to be raised or propagated, there are a few special points to note. They should not be grown in an organic medium, such as peat-based or soil-based compost, but in an inert one, which will enable the plants to make the transfer to water culture easily. If remnants of the organic medium are still adhering to the plant, these can easily rot, or provide the germ of infection. If the plant has been propagated in an organic medium, every trace of soil must be washed off before the plant is transferred to hydroculture. The greater the age of the plant, the more

Dieffenbachia in hydroculture

extensive the network of roots, and so the more difficult it will be to wash off the organic growing medium.

Choose pellets of hydrocultural growing medium which are a suitable size for propagation (such as fine expanded clay granules). Place these in a propagating container, and irrigate until water collects in the bottom of the container. After a short while, this is poured off. If a drainage layer of medium gravel is incorporated, it is not necessary to pour the water off. The desired seeds or cuttings (with their lower leaves removed) are now pressed down into the medium. The container is then enclosed under a polythene bag and placed in a warm, well lit position, ideally on a thermostatically regulated base. A propagator with a similar facility may also be used.

Once the young plants or seedlings have made about 10cm (4in) of growth, they can be planted individually into hydroculture pots.

Tips for raising plants

Flower-arranging material can be a useful substrate for rooting cuttings or raising plants from seed. This can be bought ready-made from a specialized dealer.

Various pellet sizes are available for raising plants for hydroculture.

However, the blocks of flower-arranging material available from florists can also be cut into pieces of the desired size with a knife.

At first, only a weak nutrient solution should be given. If the roots have been damaged when transplanting, it is advisable to allow the plants to continue to stand in water for a while.

Plants which can be raised in hydroculture, using propagation-sized pellets

Zebra plant (*Aphelandra*), kangaroo vine (*Cissus*), croton (*Codiaeum*), dieffenbachia (*Dieffenbachia*), pellionia (*Elatostema*), species of rubber plant (*Ficus* sp.), ivy (*Hedera*), wax plant (*Hoya*), myrtle (*Myrtus*), oleander (*Nerium*), passionflower (*Passiflora edulis* and other species), tradescantia (*Tradescantia*).

Ferns

Ferns are propagated not by seed, but by spores. These can be seen as grey or greyish-brown grainy, raised spots underneath the fronds of potted ferns, or ferns growing in the wild. Spores of a large number of species are available from the appropriate specialist seed suppliers. In order to propagate them, they are shaken (or carefully extracted from the packet, as the spores themselves are very fine) and placed on a warm, damp substrate. This compost should be sterile, to prevent harmful organisms from developing. Kept humid, the spores germinate in a matter of several weeks, and form green, flat growths, with male and female organs. The male gametes fertilize the female ones, and only then are the tiny new ferns produced. When they are a few centimetres high, they can be pricked out.

Meristem culture

Almost any of the living cells of a plant can be used to generate a young plant, under sterile conditions. The plant is stimulated at high temperature to produce a shoot. Small pieces of meristem tissue (the tiny growth tip only 1–2mm across) are removed with a scalpel. The clumps of cells multiply in a nutrient medium, and a few weeks later, callus (wound tissue) is formed. The tissue then differentiates, and roots are produced. When in the next stage shoots are formed, tiny plants are seen, each an identical clone of the original plant. These young plants may also have an advantage over the parent plant: they are free of any virus or bacterial diseases which may have been present. With this method of propagation, the juvenile phase of the plant may be more pronounced. This leads to more vigorous growth, and the unusual growth of thorns in some woody plants.

Specialized laboratories use meristem culture to raise many indoor plants and useful plants (including various rootstocks), for example, a number of herbaceous perennials such as bamboos, zebra grass, lilies, roses, orchids, palms, citrus, pistachio, loquat, rhododendrons, and fruit trees and bushes.

Plant propagation for children

Tomatoes must be amongst the most popular items on the summer menu, and are especially enjoyed by children. Home-grown tomatoes are delicious. Not only do we have the familiar fleshy tomatoes and the large, round beefsteak tomatoes but also the small cherry and sweet cocktail tomatoes. It is even possible to grow small, yellow, pear-shaped tomatoes.

They taste best, of course, if they come from plants you have grown yourself. Doing this is not difficult. Children enjoy sowing seeds in small containers in the depths of winter. The seeds are then placed on the windowsill, where they soon germinate. Now the children can watch them grow. The delicate little seedlings grow into strong, young plants, which are transferred to individual pots once they reach a height of about 10cm (4in). In early summer, after the last risk of frost, they are planted out into the garden, or into a large pot of good garden soil. This way, their fruits can ripen on a balcony or terrace. Large plants will need to be supported with stakes.

If a particularly small variety such as the cocktail tomato 'Tiny Tim' is chosen, it can even be grown in a pot on a reasonably sized windowsill, and will produce a good number of tasty fruits.

To grow a tomato tree, you will need to obtain some tree tomato (tamarillo) seeds. The fruit departments of some stores may even stock the fruit. The seeds are extracted, washed and dried, and sowing them can then begin. It is not necessary to wait until winter, as it is with the tomatoes described above. The tomato tree does in any case need time – at least two years – before it begins to flower and bear fruit and by then it will have reached the proud height of about 1m (3ft).

You will need to consider where you will overwinter a plant like that, if you decide to grow one. The ideal place is a conservatory or greenhouse. If you need to make do with a windowsill,

Growing tomatoes: the seeds are sown in small containers in winter

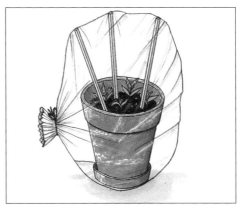

They can be watched as they germinate on the windowsill

you will have to prune the little tree back in the autumn.

Small plants can be grown very easily from orange or lemon pips. These plants grow very slowly, and in time they turn into fine little trees. They make an attractive foliage plant, even though it normally takes several years before your very own orange or lemon tree produces any flowers or fruit.

But children can find more types of plant propagation to fascinate them, apart from producing and growing plants from seeds. A bunch of forsythia, still sitting half-forgotten in a vase, can be a playground for tomorrow's gardeners. The forsythia forms so many roots after a few weeks in the vase that it is difficult to separate the stems. So take them out in good time, plant them carefully in a pot of soil-based potting compost, and water them. A fading bunch of flowers can be turned into a batch of new plants.

The ever-popular pussy willow catkins are also good candidates for rooting in a vase. You should not let too much time go by before planting them, because the glassy roots will tend to break off easily if they are allowed to grow too long.

Plants can not only be propagated by seeds and cuttings. The tropical spider plant, a popular houseplant, has its own special strategy for reproducing its kind. It puts out long, thin stems bearing tiny, ready-made plants, which already have their own roots. As these plantlets, as they are called, grow larger, the stem is bowed down until it touches the ground. There the plantlets are able to strike root, and grow on to become spider plants in their own right.

If grown as a houseplant, once the plantlets have formed some roots, the stem bearing them can be cut with a pair of scissors, and the plantlets potted into small flowerpots with propagating compost. If they are watered and placed on the windowsill, they will soon grow into new plants.

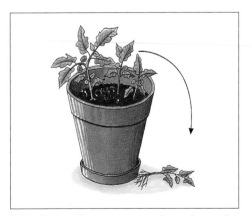

The seedlings are transplanted into the garden, or a large pot, in early summer

If the plant has enough light, red fruits will develop

Index

Plant Propagation
A WARD LOCK BOOK
First published in the UK 1997 by
Ward Lock, Wellington House, 125 Strand, London WC2R 0BB

A Cassell imprint

ISBN 0 7063 7584 X

© 1995 Falken-Verlag GmbH, 65527 Niedernhausen/Ts.

Translation: Elaine Richards in association with First Edition Translations Ltd, Cambridge
Photographs: FALKEN Archiv/hapo: 75 l.; /**Landini:** 5 above, 49 l.; /**Tessmann und Endress:**93; /**Wilhelm:** 55
Bonsai-Centrum Heidelberg, Mannheimer Straße 401, 69123 Heidelberg: 96
Monika Klock, Hamburg: 1, 22, 23 above, 94 r.
Peter Klock, Hamburg: 3, 17, 34 (3x), 35, 41 l.below
Wolfgang Redeleit, Bienenbüttel: 9, 10 above, 10 below, 11, 12 l., 12 r., 15, 16, 24, 25
Reinhard-Tierfoto, Heiligkreuzsteinach-Eiterbach: 94 l., 97
H.-J. Schwarz, Idstein: 49 r., 51, 75 r.
Gitte and Siegfried Stein, Vastdorf: 2, 4, 5 below, 13 l., 13 r., 14, 27, 63, 64, 66, 67, 68, 74 (2x), 76 (2x), 77 (2x), 78 (2x), 83, 84, 89 r.
WOLF-Geräte, Betzdorf/Sieg: 6
Drawings: FALKEN Archiv/Lünser: 18, 23 below, 36 (2x), 37 (4x), 40 r. above, 41 l. above, 44 (3x), 45, 88, 89 l.; /Scholz: 90;/ Stegeman: 19, 43 r.
All others: Ulrike Hoffmann, Bodenheim

Printed by Pozzo Gros Monti S.p.A, Italy

Distributed in the United States
by Sterling Publishers Co., Inc.
387 Park Avenue South,
New York, NY 10016-8810

Distributed in Canada by Cavendish Books Inc.
Unit 5, 801 West 1st Street
North Vancouver B.C.
Canada V7P 1A4